*The Geological Society of America*
*Memoir 69*

# GENUS *PENTREMITES* AND ITS SPECIES

BY

## J. J. GALLOWAY

AND

## HAROLD V. KASKA

*Indiana University, Bloomington, Indiana*

April 18, 1957

*Made in the United States of America*

PRINTED BY WAVERLY PRESS, INC.
BALTIMORE, MD.

PUBLISHED BY THE GEOLOGICAL SOCIETY OF AMERICA
Address all communications to the Geological Society of America
419 West 117 Street, New York 27, N. Y.

*The Memoir Series*
*of*
*The Geological Society of America*
*is made possible*
*through the bequest of*
*Richard Alexander Fullerton Penrose, Jr.*

## ACKNOWLEDGMENTS

We are most indebted to the late Dr. Clyde A. Malott for making available his extensive collection of many hundreds of specimens of *Pentremites* from many formations in the Middle and Upper Mississippian for study and identification. We also are indebted to Prof. Courtney Werner of Washington University, and Prof. R. C. Gutschick, of the University of Notre Dame, for the gift and loan of specimens, and to James E. Reeves for preliminary study of a large group of specimens from the Golconda formation at Shoals, Indiana. Karl W. Frielinghausen identified many species of the Malott collection. George R. Ringer, photographer of the Geology Department of Indiana University, photographed specimens and original figures. The drawings were made and the photographs retouched by Arthur Pinsak, Mrs. Mary Gillim, and J. J. Galloway; most drawings were made from specimens in the Paleontological Collections of Indiana University. Alan Horowitz has carefully read the entire manuscript and checked identifications and bibliography.

We are indebted to Dr. R. S. Bassler and Dr. G. A. Cooper, of the U. S. National Museum, for making the collection of *Pentremites* in the museum available for study. We are also under obligation to Dr. Charles F. Deiss, for making available many of the facilities of the Department of Geology, and to the Graduate School of Indiana University for a grant in aid of the research.

# CONTENTS

| | Page |
|---|---|
| ABSTRACT | 1 |
| PART 1. THE GENUS *Pentremites* | 3 |
| Introduction | 3 |
| Historical review and nomenclature of *Pentremites* | 5 |
| Definition of a pentremite | 8 |
| Structure of *Pentremites* | 8 |
| Glossary of Terms for *Pentremites* | 11 |
| Characteristics of *Pentremites* and their taxonomic values | 12 |
| General discussion | 12 |
| Most important characters | 13 |
| Length to width ratio; L/W | 13 |
| Vault to pelvis ratio; V/P | 13 |
| Pelvic or basal angle | 14 |
| Convexity or concavity of ambulacra | 14 |
| Concavity of interambulacra | 15 |
| Height of ambulacral rim | 15 |
| Extension of the deltoids | 15 |
| Flare of the deltoids | 16 |
| Profile of the edges of the pelvis | 16 |
| Nodes on the basal plates and radials | 16 |
| Shape of the sides of the vault and vault profile | 16 |
| Position of greatest width | 17 |
| General shape of the specimen | 17 |
| Size of the specimen | 17 |
| Secondary characters | 17 |
| Width of the ambulacra | 17 |
| Length of the ambulacra | 18 |
| Number of lateral grooves in 3 mm | 18 |
| Slope of lateral grooves | 18 |
| Finer structures of the ambulacra | 18 |
| Angle of the vault | 20 |
| Angle between pelvis and vault | 20 |
| Shape of the summit | 20 |
| Width of the summit | 20 |
| Length of the deltoids | 20 |
| Ontogenetic stages | 21 |
| Negligible characters | 21 |
| Supposed dwarfing | 21 |
| Total number of transverse grooves | 21 |
| Shape and size of lateral pores | 22 |
| Width of lancet and side plates | 22 |
| Median groove of the ambulacrum | 22 |
| Shape and size of the mouth | 22 |
| Shape and size of the spiracles | 22 |
| Basal suture of the deltoids | 23 |
| Shape of the basal plates | 23 |
| Radial pelvic ridges | 23 |
| Length of the calyx stem | 23 |
| Diameter of the stem | 23 |
| Striations of the plates | 24 |

                                                                                Page
        Abnormalities.....................................................    24
        Summit plates and ambulacral covering plates......................    24
        Pinnules..........................................................    24
        Hydrospires.......................................................    24
    Nonstructural features................................................    25
        Geographic occurrence.............................................    25
        Stratigraphic occurrence..........................................    25
        Abundance.........................................................    25
    Geologic history of *Pentremites*.....................................    25
    Ontogenic development of *Pentremites*................................    27
    Ancestry of *Pentremites*.............................................    30
    Phylogeny of *Pentremites*............................................    31
    Ecology of *Pentremites*..............................................    32
    Fossilization of *Pentremites*........................................    33
PART 2. SYSTEMATIC DESCRIPTIONS OF GENUS AND SPECIES.......................    37
    Classification of the genus *Pentremites*.............................    37
    Guide for description of species of *Pentremites*.....................    38
    Groups of species of *Pentremites*....................................    38
    Key to groups of *Pentremites*........................................    39
    Key to species of *Pentremites conoideus* Group.......................    39
    Descriptions of species of *Pentremites conoideus* group..............    40
    Key to species of *Pentremites godoni* group..........................    45
    Descriptions of species of *Pentremites godoni* group.................    45
    Key to species of *Pentremites pyriformis* group......................    51
    Description of species of *Pentremites pyriformis* group..............    51
    Key to species of *Pentremites sulcatus* group........................    61
    Description of species of *Pentremites sulcatus* group................    62
    Collecting localities.................................................    75
    Check list of species of *Pentremites*................................    76
BIBLIOGRAPHY..............................................................    81
EXPLANATION OF PLATES.....................................................    85
INDEX.....................................................................   101

## ILLUSTRATIONS
### PLATES

Plate                                                                Facing page
1. Ambulacra of *Pentremites*............................................. 86
2. *Pentremites conoideus* group.......................................... 87
3. *Pentremites godoni* group............................................. 88
4. *Pentremites pyriformis* group......................................... 89
5. *Pentremites pyriformis* and *Pentremites sulcatus* groups............. 90
6. *Pentremites sulcatus* group........................................... 92
7. *Pentremites sulcatus* group........................................... 93
8. *Pentremites sulcatus* group........................................... 94
9. *Pentremites sulcatus* group........................................... 95
10. *Pentremites sulcatus* group.......................................... 96
11. Ontogenetic series of *Pentremites*................................... 97
12. Ontogenetic series of *Pentremites*................................... 98
13. Structures and fossilization of *Pentremites*........................ 99

### FIGURES

Figure                                                                    Page
1. Geologic occurrence and phylogeny of the groups of *Pentremites*........ 28
2. Geologic occurrence and phylogeny of the *Pentremites conoideus* group.. 40
3. Geologic occurrence and phylogeny of the *Pentremites godoni* group..... 46
4. Geologic occurrence and phylogeny of the *Pentremites pyriformis* group. 52
5. Geologic occurrence and phylogeny of the *Pentremites sulcatus* group... 62

### TABLES

Table                                                                     Page
1. Tabulation of Meramec-Chester classifications used in Indiana, Kentucky, and Illinois...  9
2. Main characters of the *Pentremites conoideus* group of species......... 41
3. Main characters of the *Pentremites godoni* group of species............ 47
4. Main characters of the *Pentremites pyriformis* group of species........ 53
5. Main characters of the *Pentremites sulcatus* group of species.......... 63

# ABSTRACT

About 133 species of blastoids have been assigned to the genus *Pentremites*, of which 63 are here recognized as valid species or varieties; 30 are synonyms or unrecognizable, 10 are *nomina nuda*, and 30 belong to other genera. They range from Lower Mississippian to Late Mississippian, possibly to Lower Pennsylvanian, reaching their acme in number and variety in the lower and middle Chester. Their structures are defined and evaluated, and the valid species are redescribed and figured. The evolution of the genus and species are attempted, and the ranges and phylogenies of the species given. All species are analyzed and described after a standard plan, and keys are formulated, so that the species can be more readily and reliably identified and geologic horizons thereby determined. Three new species are proposed: *Pentremites gutschicki*, *Pentremites halli*, and *Pentremites malotti*.

1

# PART 1. THE GENUS *PENTREMITES*

## INTRODUCTION

Specimens of *Pentremites* occur in abundance in the Middle and Upper Mississippian and rarely in the Lower Mississippian. One hundred and three species which belong in the genus *Pentremites* proper and 30 forms called *Pentremites* which belong in other genera have been named. Each author has used his own criteria for distinguishing species but most authors have been unable to recognize species of other authors. The present writers recognize about 62 American species and varieties and 1 doubtful species from Europe.

The authors have studied the published works of nearly all other students of *Pentremites* and specimens of most of the species and varieties, including the collections of Lyon, Meek and Worthen, Hambach, Ulrich, Mather, and Butts, now in the U. S. National Museum. The genus is in such a state of confusion that it is almost impossible for even an experienced paleontologist to identify a species. Differences in abilities, training, and experience, differences in criteria used, lack of descriptions, poor figures, scattered and obscure literature, ignoring of previous work, and failure to recognize ontogenetic stages and individual variations, have made the identification of species of *Pentremites* and their use in identifying geologic formations almost impossible. The writers have not tried to revolutionize the identification and classification of the pentremites but have evaluated the taxonomic characters and synthesized the work of others to make it more useful.

The need for monographic treatment of all species of the genus, applying the same criteria for distinguishing all species, and the need for making the species more useful in identifying and correlating stratigraphic formations, have prompted the present study. Attention is here directed toward identification of species and their use in stratigraphy, rather than to anatomical, physiological, ecological, and teratological problems, for which see the works of Hambach, and of Etheridge and Carpenter.

Thirty-five species which still belong to the genus *Pentremites* were erected from 1825 to 1886. The culmination of the study of the structure and classification of the *Pentremites* by Hambach came in 1880 and 1884, and the great work of Etheridge and Carpenter was published in 1886. Etheridge and Carpenter were distressed with the large number of species which had been erected up to that time (Hambach 9, Lyon 5, Hall 4, Troost 3, Meek 2, Say 2, Shumard 2, and Defrance, Schlotheim, Etheridge and Carpenter, Conrad, Meek and Worthen, Owen, Roemer, and Swallow, 1 each). They say (1886, p. 128),

"The number of so-called species of *Pentremites* would be very greatly reduced if they were all subjected to a critical examination which was based on the comparison of a large amount of material . . . we may not unfairly conclude that a revision by competent hands of the host of different species of *Pentremites* which have been described in America would reduce a great proportion of them to the rank of synonyms."

No new species were named for the next 14 years. In spite of the plethora of old species which could not be distinguished, there were added 52 new species and varieties from 1900 to 1945 (Ulrich 30, Hambach 7, Rowley 4, Clark 3, Haas 2, Weller 2, Schuchert 1, Butts 1, Greger 1, and Easton 1; Ulrich also proposed (1917, p. 147, 105,

226) 9 *nomina nuda*. In that time most of the old species were ignored, and only about a dozen species of the 35 original species have been identified again in the past 70 years. It is improbable that even a dozen of the 52 species erected since 1900 will be recognized again unless something is done to bring descriptions, figures, evaluations, and keys into one synthetic work.

The present writers have made a serious effort to distinguish all described and figured species and varieties. Many more named forms probably should have been placed on synonymy, instead of recognizing the 63 forms which are distinguished. There are about 40 readily recognizable species.

The writers have recognized only three new species, *P. gutschicki*, *P. halli*, and *P. malotti*, each readily distinguished from closely related forms occurring with them. Other new species will be found as new collections are made and the genus is studied further. It would be possible to erect several hundred new species and varieties by insisting on precise numerical measurements for all the 42 or more structural characters employed. Species of *Pentremites* vary to about the same degree as do other kinds of organisms, but the variants occur with the typical forms of the species, and there is no practical value in naming them. Drawing specific lines too closely causes extreme difficulty and loss of time in identification and doubt concerning the accuracy of identifications. Well-characterized varieties are of value in correlation of strata, but most variations occur with the species and are therefore of no additional value in correlation. It is already difficult to identify a species of *Pentremites*, as anyone who attempts it will soon realize.

Many published figures differ somewhat from the type figures of the species, *e.g.*, the figures by Rowley (1901–1904), by Ulrich (1917), and by Haas (1945). The amount of variation allowable is a matter of judgment, and the tendency is to divide species finer and finer. Some paleontologists allow more variation within a species, *e.g.*, *P. welleri* Ulrich (1917, Pl. 6) than between different species, *e.g.*, *P. welleri* Ulrich and *P. abruptus* Ulrich (1917, Pl. 6). It seems better to allow some latitude in variation of characters than to make new but unrecognizable species and varieties.

The writers recognize four groups of species of *Pentremites*, which best show the geologic occurrence and phylogeny of the genus. The groups also make possible analytical tables and keys to the species. New generic or subgeneric names for the groups are not proposed, since these additions would complicate identification and add nothing to the knowledge or usefulness of the genus. Ten or 12 groups could be made, as is actually done in constructing the phylogenetic trees and in making the keys to species. But none of such groups should be made into new taxonomic groups or given taxonomic names.

The purpose of the present paper is to make identification and use of *Pentremites* more accessible to paleontologists, more useful to stratigraphers, and more alluring to amateur collectors.

The important early works of Parkinson, Say, Roemer, Shumard, Lyon, Hambach, Rowley, and others are in few libraries and are difficult to find. Even the work of Ulrich (1917), in which the most species are erected for a single work, is out of print. It was therefore necessary to reproduce the original type figures of all species to make it possible to identify them with a minimum of effort.

The identification of species and varieties of *Pentremites* calls for all the ability, study, and experience that the student of the group can command. It also requires the ability to discern and judge soundly the relative values of the multitude of variables—structures, ontogenetic stages, changes in time, adaptations to ecological conditions, individual variations—with which the student must deal.

Students of *Pentremites*, of whom there have been more than a score, have labored under at least two great difficulties: (1) each has used his own criteria for distinguishing species: (2) they have ignored almost entirely other authors' species, probably because they could not recognize them.

## HISTORICAL REVIEW AND NOMENCLATURE OF *PENTREMITES*

The study of the genus *Pentremites* in America closely coincides with the development of systematic paleontology in America. Scientists early noted these attractive little fossils in the Lower Carboniferous rocks. Since the first published description and illustration of the genus by Parkinson (1808, p. 235), much has been written concerning their structure, physiology, classification, and stratigraphic occurrence. Parkinsons' remarks concerning the fossil are quoted below:

"The nature of the calcareous fossil, represented Plate XIII. Fig. 36 and 37, is very ambiguous. I was favoured with it by Dr. Woodhouse, of Philadelphia, who, avowing his inability to ascertain any thing respecting its original mode of existence, informs me, that it was obtained from Kentucky, where similar bodies are frequently found of a larger size, and that they are there considered as a species of petrified nuts.

"This fossil is somewhat of a conical roundish form, the centre of its base terminating in a small round projection pierced in its middle, with a little opening into the centre of the fossil: from this projection the base extends nearly horizontally to five prominent points, between each of which exists a shallow depression. At the apex of the cone five small openings are placed at the angles, formed by the meeting of the lines, which bound five long triangular surfaces, which, commencing at the summit of the fossil, are disposed tapering, down the sides, and terminate in the projecting points which are placed round the base. Along the middle of each of these surfaces, a grooved line passes, from which upwards of forty minute processes on each side, pass to the lines which bound these surfaces at their sides. [Parkinson's figures are reproduced on Pl. 3, figs. 11a, 11b.]

The opinion which I formed on the first view of this fossil was, that it belonged to some animal approximating to the encrinus. The central projection at its base has, however, suffered so much by friction as not, I believe, to show its original surface; and hence it is impossible to determine at present any thing respecting the substance with which it was connected at this part, or the kind of articulation which was here employed. It however, very nearly resembles the smaller *modiolus*, from Ireland, which is represented in Plate XVII." [The specimen figured on Plate XVII is the calyx of a crinoid.]

No further work was done with *Pentremites* for the next decade, except that Mitchill (*in* Cuvier, 1818) briefly noted the occurrence of forms now called *Pentremites* but did not advance knowledge of the genus to any great degree.

Defrance (1819, p. 467) first attempted to name properly the fossil described by Parkinson. He wrote in the *Dictionnaire* that he had named the fossil, seen in Parkinson's *Organic Remains, encrina Godonii* in honor of Godon. Weller (1920, p. 320) stated that he did not believe that Defrance had access to actual fossil material, but it is apparent that he did, for he writes:

"J'ai reçu de l'Amerique septentrionale de jolie corps, que l'on trouve dans le Genesee, et qui pourroient dependre de la famille des encrines. . . . Je lui ai donné provisoirement le nom d'encrine de Godon, *encrina Godonii*. On en voit la figure dans Parkinson tab. 13, fig. 36 et 37."

No other figures are known to have been made by Defrance, so that the valid specific name created by him must be applied to Parkinson's illustrations.

Von Schlotheim (1820) probably, as Weller noted (1920, p. 320), used Parkinson's illustrations, applying to them the name *Encrites florealis*. The present writers agree with Weller's conclusions that *Godonii* and *florealis* are strictly synonyms. Parkinson's original specimen is the holotype and his illustrations are the type figures.

Thomas Say (1820a, p. 35) credited Parkinson as the first author to describe and figure *Pentremites* under the name "Kentucky Asterial fossil". He stated that he had examined numerous specimens in the collection of the Academy of Natural Sciences collected by Samuel Hazard, near Huntsville, Alabama, and was convinced that they were not members of the genus *Encrinus* but were actually a separate genus which he termed (p. 36) *Genus Pentremite*. He recognized three forms of *Pentremites* based on the shape of the pelvis but did not name them. Say said (1820a, p. 38) that he had been informed that Mitchill had noticed and figured *Pentremites* as an "*Echinus of the family* (genus) *Galerite*" and also as an "*asterite*".

Say's next contribution (1825, p. 292) to the study of *Pentremites* was to propose the family *Blastoidea* and recognize that it contained forms intermediate between the crinoids and echinoids. At this time, Say separated the three forms of *Pentremites* mentioned in 1920 into three species, *P. globosa*, *P. pyriformis*, and *P. florealis*. In the appended synonymy of *P. florealis* Schlotheim, Say lists the "*Kentucky Asterial fossil*" of Parkinson. Since Defrance named Parkinson's figure a year earlier than von Schlotheim, and law of priority applies, and Say should have called the fossil *Pentremites Godonii* (Defrance), now spelled *Pentremites godoni* (Defrance).

Sowerby (1826, p. 316) in commenting on Say's paper of 1825 corrected *Pentremite* to *Pentremites*, correctly crediting Say as the author of the genus. He refers to "*Description of a new Species of Pentremites*" and states, "The almost anomalous form and singular structure of the bodies distinguished by Mr. Say by the name of *Pentremite* (*Pentremites*) has been the cause that some attention has also been given to them in this country."

Although not directly stated in the article, it seems that Sowerby realized that pentremite, when written with a small initial letter, is a correct English term, and the correct Latin form is *Pentremites*. The name *Godonii* has also been changed to *godoni* in accordance with the Rules of Nomenclature. (Art. 14, C, "If the name is a modern patronym, the genitive is always formed by adding to the exact and complete name, an *i*.")

Goldfuss (1826, p. 150) was dissatisfied with the etymological constructions of both Say and Sowerby. He therefore changed the name of the genus to *Pentatremites*, but later workers have consistently rejected this proposal, favoring instead the form used by Sowerby, *Pentremites*, and regarded by the present writers as correct.

Roemer (1851, p. 324) further emended *Pentremites* to *Pentatrematites*, stating that it was derived from the Greek *penta*, five; and *trema*, foramen, a hole. It is not, as Miller (1889, p. 267) stated, derived from "*pente*, five; *remos*, a board or plate". The writers believe Roemer's is the correct derivation but disagree as have most others with the necessity for lengthening the name *Pentremites*, since it is perfectly proper to shorten Latinized combining forms. The writers regard the *ites* as having been derived from the Greek *lithos*, *lites*, *ites*, stone. Roemer (1851, p. 324) proposed a classification of the pentremites, actually blastoids, based on the shape of the calyx and the size relations of the ambulacra. It consisted of the following four groups.

(1) Floreales, type; *P. godoni* (Defrance) (equals *Pentremites*)

(2) Elliptici, type; *P. ellipticus* Sowerby (equals *Orbitremites*)

(3) Truncati, type; *P. pailleti* de Verneuil (equals *Pentremitidea*)

(4) Clavati, type; *P. reinwardtii* Troost (equals *Troostocrinus*)

Most of the "pentremites" classified thus by Roemer were mainly European forms which have since been relegated to other blastoid genera, such as *Codaster*, *Nucleocrinus*, *Orbitremites*, *Pentremitidea*, and *Troostocrinus*. However *P. sulcatus* Roemer is a true pentremite and the type of the *P. sulcatus* group.

Lyon (1857, p. 468) proposed that three small basal pieces below the usual three basal plates be added to the generic formula or description of *Pentremites*, making the number of plates in a calyx 16 instead of 13. He also described a new species, *P. obesus*. Hall (1858, p. 484, 607, 690) published descriptions and excellent figures of species of *Pentremites*, some of which have since been put in other blastoid genera. Shumard (1858, p. 238) was also interested in this comparatively new group of fossils. He published a synonymy with references and commented on the occurrence of the small calcareous plates covering the mouth and spiracles. These have frequently been referred to in the literature but occur on few specimens.

One of the most important of the American workers at this time was Hambach (1880, p. 145; 1884a, p. 537; 1884b, p. 548) who, before the end of the century, had published 9 new species and written extensively on the structure and classification of the pentremites.

In 1886 Etheridge and Carpenter published their *Catalogue of the Blastoidea* which summarizes all the knowledge of the blastoids existing then and contains material of great value to the student of *Pentremites*. Complete descriptions and synonymies characterize the volume, as well as historical and anatomical discussions. Etheridge and Carpenter proposed a new classification of the blastoids based on their earlier work on the class. They divided the blastoids into two orders, *Regulares* and *Irregulares*, and six families, *Pentremitidae*, *Troostoblastidae*, *Nucleoblastidae*, *Granatoblastidae*, *Codastridae*, and *Astrocrinidae*; the latter is the only family listed under the *Irregulars*. The *Pentremitidae* contained the three genera *Pentremites* Say, *Pentremitidea* d'Orbigny, and *Mesoblastus* Etheridge and Carpenter. They state (p. 156) that they were "... convinced that many synonyms exist" and that the genus was "... essentially an American one".

An interesting feature concerned with a historical review of the pentremites is the scientific feuds in which Dr. Hambach became involved (1884a, p. 537; 1903, p. 36) first with Carpenter (1881), and later with Rowley (1904, p. 192).

The interest in the genus died down after the *Catalogue of the Blastoidea* was released but was not extinguished, for many small papers were published. Hambach (1903), in the major work between 1886 and 1917, proposed a new classification of the blastoids. He divided them into the orders *Regulares* and *Irregulares*, the former including the families *Pentremidae* and *Codasteridae* and the latter including the families *Olivanidae* and *Eleutheroblastidae*. *Pentremites* is one of the eight genera listed under the *Pentremidae*. At this time he also proposed new species, several of which the present authors consider varieties of *P. godoni*. Modern classification considers *Pentremites* a genus of the order *Eublastoidea*.

Dr. E. O. Ulrich became interested in the pentremites while gathering material for his study of the Chester Series of Kentucky. He hoped that they would prove of great assistance in unraveling the stratigraphic relations of the complex alternation of limestone, sandstones, and shales which characterize this series. Ulrich was the most prolific contributor to the species of *Pentremites*, having named 30 species and varieties plus 9 *nomina nuda*. Three species and a variety were described in 1905 and the others in 1917. The writers recognize most of these, but in certain cases, Ulrich included in his species specimens which are not conspecific with the type. In the case of *P. welleri* (1917, Pl. 6, figs. 15–26), there are at least two other species— *i.e.*, Figure 21 is *P. buttsi*, and Figure 22 is *P. altus*. Ulrich frequently divided his species very closely and yet included several species under one name. His work is of great value, particularly because of the excellent figures and the keenness of some of his observations. He has the first to recognize that shape of the ambulacra is important in classification and mentioned it in almost all his descriptions. It seems established, however, that *P. planus* Ulrich is definitely the same as *P. godoni* (Defrance), and practically all later workers have also taken this view.

Weller (1920) described and figured two new species of *Pentremites* and gave a good discussion of the synonymy of *P. godoni*. Greger (1934) made a decidedly useful contribution to the study of the blastoids in publishing his *Bibliographic Index of North American Species of the Eublastoidea*. This publication was of great assistance in locating American references, practically all of which the writers have examined.

The classification of Mississippian rocks used in this paper closely follows the *Correlation of the Mississippian Formations of North America* by Weller *et al.* (1948, p. 91–188). The classification of the Middle and Upper Mississippian as used by various authors is given in Table 1.

## DEFINITION OF A PENTREMITE

A pentremite is an extinct, middle Paleozoic, marine, budlike organism, belonging to the class Blastoidea of the phylum Echinodermata. It was attached to the sea bottom by a short, round, jointed stem, 1–3 inches long. It consists of 13 plates, five food-gathering areas, a central mouth, and five summit pores. In life there were pinnules on the ambulacra.

## STRUCTURE OF *PENTREMITES*

The head or calyx is pentamerally and bilaterally symmetrical and consists of three basal plates, as in some crinoids, five forked plates or radials, alternating with the basals and containing the ambulacra, and five quadrangular or deltoid plates, alternating with the radials. The plates are united at the sutures, on both sides of which growth took place. The stem columnals grew from the center of the base of the calyx and were cut off as growth was completed, leaving a flexible suture. Thus the oldest structures are the upper tips of the deltoids, the lower points of the ambulacra, the lower part of the fork of the radials, and the lower parts of the basal plates. The

TABLE 1.—*Tabulation of Meramec-Chester Classifications Used in Indiana, Kentucky, and Illinois*

| Weller *et al.* (1948) | Ulrich (1905; 1917) | Weller and Sutton (1940) | Lyon (1860) |
|---|---|---|---|
| | | | "KASKASKIA" GROUP OF HALL |
| Kinkaid<br>Degonia<br>Clore<br>Palestine<br>Menard<br>Waltersburg<br>Vienna<br>Tar Springs | Clore<br>Palestine<br>Menard<br>Tar Springs (BIRDSVILLE) | Kinkaid<br>Degonia<br>Clore<br>Palestine<br>Menard<br>Waltersburg<br>Vienna<br>Tar Springs | |
| Glen Dean<br>Hardinsburg<br>Golconda<br>Cypress | Glen Dean<br>Hardinsburg<br>Golconda<br>Cypress | Glen Dean<br>Hardinsburg<br>Golconda<br>Cypress (OKAW) | 3rd limestone (Glen Dean)<br>3rd sandstone<br>2nd limestone (Golconda)<br>2nd sandstone |
| Beech Creek<br>Elwren<br>Reelsville<br>Sample<br>Beaver Bend<br>Mooretown<br>Paoli<br>Aux Vases | Gasper (MONTE SANO)<br>Aux Vases | Paint Creek<br>Bethel (RENAULT)<br>Downeys Bluff<br>Shetlerville<br>Aux Vases | 1st limestone (Lower Chester)<br>1st sandstone |
| | | | MILLSTONE—GRIT |
| Ste. Genevieve | Ohara<br>Rosiclare<br>Fredonia | Hoffner<br>Levias<br>Rosiclare<br>Fredonia | Cavernous limestone |
| St. Louis<br>Salem<br>Harrodsburg | St. Louis<br>Spergen<br>Warsaw | St. Louis<br>Salem<br>Warsaw | Middle limestone |

UPPER · MIDDLE · LOWER — CHESTERIAN

MERAMECIAN

MIDDLE AND UPPER MISSISSIPPIAN

early parts of the 13 plates became greatly separated by growth along the sutures, and the ambulacrum grew at the upper end and the sides, not at the lower end (Pl. 13, fig. 1).

The basals consist of a small quadrangular plate, directed toward the right anterior interambulacrum, and two larger, five-sided plates, each formed by the ankylosis of two plates equal in size and shape to the smallest plate. Growth of the smaller plate and the two larger plates was peripheral, not along the ankylosed sutures, for the union of the two component plates took place by the end of the embryonic stage. The writers do not agree with Lyon (1857, p. 469) that there were three small additional plates next to the stem attachment. Many species with nearly flat base, such as *P. godoni*, *P. platybasis*, and *P. tulipaformis*, have a triangular depression about the stem cicatrix. The sutures between the three ordinary basal plates extend through the corners of the small triangle to the center of the stem, but there are no sutures in the triangular depression. Workers since Lyon have not recognized the three small basal plates of Lyon.

The ambulacra are composed of transverse ridges and grooves, which alternate on the two sides of the ambulacra and meet at the median groove. There are from 6 to 12 transverse grooves in a space of 3 mm, averaging 8 or 9; there are fewest at the lower end of the ambulacrum and most at the upper or wider end. Transverse ridges and grooves meet at an upward angle up to 90° in the early or lower part and become transverse in the upper part of the ambulacra. The inner ends of the two series of ridges form a plate which is pointed downward and embraces about half of the width of the ambulacrum; it is called the "lancet plate" (Pl. 1, figs. 3, 9–12). The outer, small, transverse plates are called "side plates" or "poral pieces", since the outer ends bear pores leading to tubes below termed the hydrospires (Pl. 1, figs. 6–9). The side plates do not cover the lancet plate, as is true for *Pentremitidea*.

The summit of the pentremites consists of the central stellate mouth, or peristome, in which the median grooves of the ambulacra end. Between the tips of the deltoids and the mouth there are five oval openings termed the spiracles. Four of these are divided by the radial extension of the deltoid plates into two tubes at a depth of 1 or 2 mm (Pl. 2, fig. 4b). One of the spiracles is larger and termed the anal opening. It consists of three tubes a little below the common opening (Pl. 13, fig. 5b). In life there may have been an apparatus extending upward from the mouth and spiracles, but it is not preserved in the specimens examined (Hambach, 1903, p. 14, figs. 6, 7).

The ambulacra were covered in life by pinnules, as shown by Roemer (1851, Pl. 5, figs. 7 a, b, c), by Etheridge and Carpenter (1886, Pl. 11, figs. 16–17), and by Butts (1926, Pl. 59, fig. 4). None of the writers' specimens show the pinnules, nor were any pinnule pieces found in the rock residues. The transverse ridges and grooves have finer transverse ridges and grooves, which turn around the inner ends of the ridges and become transverse to the median groove (Pl. 1, figs. 1–6).

Near the outer ends of the transverse grooves near the middle of the side plates, the best-preserved specimens have small pits, which may have been for the attachment of the pinnules (Pl. 1, figs. 3, 5, 7, 9, 10, 11). The pits should not be confused with the hydrospire pores, at the outer edges of the side plates, which are shown well only on weathered specimens.

Below the poral pieces there are flattened calcareous tubes, the hydrospires, into which the marginal pores open. There are from four to nine hydrospires on each side of each ambulacrum. The hydrospires can be seen beneath the ambulacra in some partially broken specimens and can be observed in a few thin sections across the calyx of specimens filled with fine mud (Pl. 7, fig. 9).

*Pentremites* differs from its nearest relatives, *Pentremitidea* and *Mesoblastus*, in not having the lancet plate covered by the side plates.

## GLOSSARY OF TERMS FOR *PENTREMITES*

*Ambulacra.* The structures in the radial sinus, consisting of the lancet plates exposed along the food or median groove, the side plates, the hydrospire pores, and the outer side plates.

*Ambulacral flange.* The elevated rim of the radial and deltoids around the ambulacrum, as in *P. spicatus.*

*Ambulacral ratio.* The ratio of the length of the ambulacra (measured parallel to the axis of the calyx from the fork to a point midway between the tips of the deltoid plates) to the width of the ambulacra (measured between the tips of the deltoid plates).

*Ambulacral rim.* The rim of the radials and deltoids bounding the two sides and lower end of the ambulacra, as in *P. pyriformis.*

*Anal pore.* The largest spiracle.

*Apex.* The highest point on the calyx, either at the mouth or tips of the deltoids.

*Basal angle.* Same as pelvic angle. The angle made by the sides of the pelvis, measured from the lower ends of two ambulacra which are 144° apart, to the center of the base. It is the angle made by the lower edges of the radial plates and edges of the basal plates, omitting most of the stem.

*Basal periphery.* The outline of the calyx as seen from the base.

*Basal plates.* One quadragonal and two pentagonal plates attached to the columnals below and the lower edges of the radial plates above.

*Basalia.* The three basal plates.

*Base.* The pelvis.

*Brachioles.* Joined arms attached to the side plates. Same as pinnules.

*Calyx.* The entire head, including all structures associated with the pelvis, vault, and summit. Sometimes less appropriately called the theca.

*Columnals.* The disk-shaped plates of the columnar stem. Rarely preserved.

*Deltoid plates.* The five rhomboid plates situated on the upper tips of the radial plates. Interradial plates.

*Dorsad.* Toward the base or stem.

*Dorsal region.* The pelvis below the ambulacra.

*Ephebic.* Structurally adult individual, which may be smaller than normal (anaephebic), of normal size (metaephebic), or unusually large (paraephebic), merging into the gerontic or senile stage.

*Forked plates.* Radial plates, embracing the ambulacra.

*Gerontic.* Old age, characterized by large size, loss of structures, or development of unusual structures.

*Hydrospires.* Ten bundles of lamellar tubes, one bundle below and parallel to each side plate, composed of four to six tubes. Seen only in cross sections of well-preserved specimens, or in specimens with the lancet plates removed.

*Hydrospire pores.* The openings perforating the side plates or outer side plates, at the outer edges of the ambulacra, generally not exposed and seen in weathered specimens. Lateral pores.

*Interambulacral areas.* Areas of the vault between the ambulacra and above the pelvis.

*Interambulacral margins.* The line, rim, or flange bordering the ambulacra at the edges of the interambulacral areas.

*Lancet plates.* The long narrow plates situated in the fork of the radial plates and traversed by a median groove. The lancet plates are partly covered by the side plates.

*Lateral grooves.* Transverse grooves across the lancet plate and side plates. They meet at an upward angle at the lower end of the ambulacrum and are in straight lines at the upper end. They are coarsest at the lower end and finest at the upper end.

*Lateral pores*. Same as hydrospire pores.

*Length*. Greatest dimension between summit or ends of deltoids and bottom of stem to which columnals were attached. Length-width ratio is length divided by width, L over W or L/W.

*Length of ambulacra*. Distance from upper end of ambulacra, below spiracles to lower pointed ends of ambulacra, measured parallel to axis of calyx.

*Mouth*. Stellate opening at the summit.

*Neanic*. Youthful stage in an individual's life; immature growth stage.

*Nepionic*. Stage between the embryonic and neanic stages.

*Outer side plates*. Row of plates situated between the outer edges of the side plates and the inner edge of the radial plates; in rare species; rarely seen.

*Pelvic angle*. Same as basal angle.

*Pelvis*. The portion of the calyx below the forks of the radial plates; dorsum, base. (Hambach included only the basal plates in the pelvis.)

*Pinnules*. The appendages, rarely preserved, which were attached to the side plates. Also called brachioles.

*Pits*. Small, round depressions on transverse grooves (Pl. 1, figs. 7–11).

*Poral pieces*. Transverse ridges across the lancet plates and side plates, with pores at the outer ends.

*Pores*. Round or oval openings at the outer ends of the lateral grooves, seen only in weathered specimens. Hydrospire pores.

*Profile*. The outline of the calyx, or part of it, as seen in side view.

*Pyriform*. Pear-shaped.

*Radial plates*. The five forked plates resting on the basals.

*Radial sinuses*. Indentations in the radial plates enclosing the ambulacra.

*Side plates*. The elongated plates between the lateral edges of the lancet plate and the inner edge of the radial plate, or the outer side plates in rare species, bearing the hydrospire pores.

*Spiracles*. The five openings at the upper ends of the deltoid plates, the largest of which is the anal pore, which is composed of the joining of three tubes. The four smaller spiracles are each composed of the union of two tubes.

*Stem*. The round lower end of the pelvis and of the basal plates, to which the columnals were attached. Also the column composed of columnals.

*Summit*. The portion of the calyx above or between the tips of the deltoid plates, at the mouth.

*Suture*. The junction of one plate with another.

*Theca*. The calyx.

*Transverse ridges and grooves*. Ridges and grooves extending from the median groove of the ambulacrum, where they alternate, across the right and left sides of the lancet plate and across the side plates, each ridge ending in a hydrospire pore. The "poral pieces" of some authors, which are neither separate plates nor divided by sutures from adjacent ridges. Also referred to as lateral ridges and lateral grooves.

*Vault*. The portion of the calyx above the pelvis, including the ambulacra, upper part of the radials, the deltoids, and the summit; the venter.

*Vault-pelvis ratio*. The ratio expressed by the height of the vault divided by the heights of the pelvis, both measured parallel to the axis of the calyx. V over P or V/P.

*Ventrad*. Toward the mouth.

*Ventral region*. The vault.

*Width*. The greatest transverse diameter, generally at the base of the ambulacra.

*Width of ambulacra*. Distance measured between the deltoid plates, or greatest transverse dimension.

## CHARACTERISTICS OF *PENTREMITES* AND THEIR TAXONOMIC VALUES

### GENERAL DISCUSSION

Species of *Pentremites* are identified by comparing a specimen with the type figure and other authentic figures and with descriptions of the species. The holotype or cotypes are rarely available for comparison. All the comparisons are made from the outside of the specimen, for the internal structures are few, hydrospire tubes and under side plates only, and are not ordinarily available. Comparisons are made structure by structure, rather than by general comparison. Most descriptions are inadequate for recognition of a species; those of Hall, Hambach, and Weller are the best. Ulrich named the most species, 30, plus 9 *nomina nuda*, but not one was actually described; he does call attention to the principal features in the explanation of figures, and most of the figures are very good. Clark (1917, p. 362) advocated the adjustment of measurements to a standard height of 20 mm; such adjustment is not advantageous. Burma (1949, p. 95) advocates the manipulation of eight measurements, in a long involved mathematical calculation; he does not say what results are obtained by mathematical comparison of two groups from *P. godoni*, which were already separated by inspection. He uses only eight characters and the length and width of structures, whereas at least 42 different structural characters, as well as geographic and stratigraphic occurrence and abundance, may be considered in identifying a species. Burma's method would not be worth the enormous effort required, even if the results were of much value.

Three groups of characters of different taxonomic importance are given; most important, secondary, and negligible. In the following discussion the attempt is made to evaluate the importance of each of the characters which previous students have described and several which are introduced in this paper, so that other students may have a better basis for description and identification than has heretofore been possible. These characters will be discussed in order of value in identification of species.

The specimen should be oriented for measurements and illustration with an ambulacrum toward the observer. All ambulacra of a specimen seem to be alike, differing only in preservation. The axis of the specimens should be at right angles to the line of sight, and in illustration the axis should be parallel to the page.

### MOST IMPORTANT CHARACTERS

(*1*) *Length to width ratio, or L/W.*—The length-width ratio of the calyx is least in *P. godoni abbreviatus*, 0.8, greatest in *P. kirki* and *P. clavatus*, 2, and generally 1.1 to 1.5. For each species, it varies little—0.1 or 0.2. This ratio does not distinguish between short pelvis and long vault, e.g., *P. buttsi*, and long pelvis and short vault, e.g., *P. okawensis*. But it is of value in distinguishing specimens which are otherwise similar, as in determining whether a specimen belongs in *P. conoideus*, L/W 1.3, or to *P. conoideus perlongus*, L/W 1.5 (Pl. 2, figs. 8, 9, 16–18). Length and width are the first measurements made, and the length-width ratio is one of the first five most important characters. The measurements are made to the nearest millimeter by using a sliding vernier caliper. The ratio obtained is significant only to tenths.

(*2*) *Vault to pelvis ratio, or V/P.*—The length of the vault is measured from the lower points of the ambulacra to the summit of the vault, parallel to the axis of the calyx. The length of the pelvis is measured from the lower tips of the ambulacra to the end of the calyx stem, parallel to the axis of the calyx. The length of the pelvis divided into the length of the vault gives the vault-pelvis ratio or V/P. This ratio determines the shape of the calyx and is a concomitant of the pelvic angle. The earliest *Pentremites* have a very short pelvis and relatively long vault. The vault-pelvis ratio varies from 4 to 20 in the *P. conoideus* group, 2 to 20 in the *P. godoni* group, 0.7 to 3 in the *P. pyriformis* group, and 0.6 to 8 in the *P. sulcatus* group. The pelvic ratio varies greatly in species with very short pelves, as in the *P. conoideus* and *P. godoni* groups, where it varies from a quotient of 2 to 20. In species with long pelves the variation is only 0.1 or 0.2. The vault-pelvis ratio is one of the first five most important structural features of a species.

(3) *Pelvic or basal angle.*—The pelvic angle is measured with a contact goniometer or a protractor, from the lower tips of the ambulacra along the radial ridges to the center of the basal plates, not including the protuberance of the stem. The stem is longest just before a new columnal is cut off, and its downward extension should not be included in measuring the pelvic angle. The concavity of the sides of the pelvis is also neglected in the measurement of the pelvic angle. The angle of adults varies as much as 10° for species with low pelvic angles, 35°–85°, although the variance is usually within 5°. It varies as much as 30° in forms with the highest pelvic angles, 100°–180°, usually varying around 10°. Since the radial ridges, along which the pelvic angle is measured, are not opposite but are 144° apart, several measurements of the pelvic angle are advisable in order to arrive at the angle which is probably correct. Since figures of pentremites show the projection of the two sides of the pelvis, the angle as measured by a contact goniometer or protractor is presumed to be the correct one. The five sides of a pentremite are not always symmetrical, and many specimens are distorted; thus a variation of 5° to 10° is expectable on the same specimen.

Young specimens have longer pelves, therefore smaller pelvic angles, than the adults of the same species. For instance, nepionic specimens of *P. conoideus* ("*P. benedicti*" Rowley) have pelvic angles of 50° to 70°, neanic specimens of *P. conoideus* ("*P. koninckanus*" Hall) have pelvic angles of 80° to 90°, whereas the angle in the adult varies from 120° to 150° (Pl. 11, figs. 1–19). Notwithstanding the variable pelvic angle of the adult and the smaller angle of the young specimens, the pelvic angle is one of the two most reliable characters for arranging the pentremites into groups of related species and for identification of species; the other most reliable character is the convexity, flatness, or concavity of the ambulacra.

(4) *Convexity or concavity of ambulacra.*—The profile of the cross section of the ambulacra is of prime importance for specific identification. The pentremites are divisible into four groups on the basis of the basal angle and the convexity, flatness or concavity of the ambulacra. In the *P. conoideus* group the ambulacra are convex (Pl. 2, figs. 1–28); in the *P. godoni* and *P. pyriformis* groups the ambulacra are slightly convex to flat (Pl. 3, figs. 1–24; Pl. 4, figs. 1–37; Pl. 5, figs. 1–21); in the *P. sulcatus* group the ambulacra are moderately to strongly concave (Pl. 5, figs. 22–30; Pl. 6–10, 12).

All species of *Pentremites* from the Lower and Middle Mississippian (except for the neanic form, *P. praematurus* from the Ste. Genevieve limestone) have convex ambulacra (Pl. 2, figs. 1–28). In the upper part of the lower Chester, the Paint Creek formation, the ambulacra become less convex and are predominantly flat at the upper ends. In one lower to middle Chester species, *P. platybasis*, the ambulacra are slightly concave at the upper ends (Pl. 3, figs. 25–29). This feature was noted by Weller and Sutton (1940). In the Golconda formation 15 species have flat ambulacra, and 6 species have concave ambulacra; in the Glen Dean limestone 8 species have flat ambulacra, and 20 species have concave ambulacra; in the upper Chester 3 species have flat ambulacra, and 5 species have concave ambulacra.

The two sides of each ambulacrum are biconvex in most species and conspicuously biconvex in *P. conoideus*, *P. biconvexus* (suggesting the specific name), and *P. hambachi*. In some species the two sides of the ambulacra are practically flat, as in *P. godoni* (which Ulrich mistakenly renamed *P. plana*, because of the plane ambulacra). In *P. spicatus*, *P. tulipaformis*, and some other species of the *P. sulcatus* group, the two sides of the ambulacra are concave.

(5) *Concavity of interambulacra.*—At least two-thirds of the species of *Pentremites* have concave interambulacral areas, including the earliest species, and including all the species of the *P. conoideus* group and most of the species of the *P. godoni* and *P. sulcatus* groups. In many species of the *P. pyriformis* group the interambulacra become nearly flat about midway between the ends of the ambulacra. The interambulacra are convex only in the nepionic stage. In basal view the outlines of most species of the *P. conoideus* and *P. sulcatus* groups tend to be stellate, whereas in the *P. godoni* and *P. pyriformis* groups of species the outlines in basal view tend to be pentameral. The basal view is never circular, except in the nepionic stage. The degree of concavity of the interambulacra is an important specific character.

(6) *Height of ambulacral rim.*—The ambulacra are usually outlined by a slightly raised flange or rim on the radial sinus and the superadjacent deltoids. The "rim" does not refer to the vertical edge of the radials and deltoids but to the rim around the interambulacrum and bordering the ambulacrum and rising from the fork piece of the radial plate, and from the outer edges of the deltoids. There is almost no rim in the early pentremites or in the *P. conoideus* group as a whole. In the *P. godoni* group there are usually low, sharp rims, about half a millimeter high, and in most species of the *P. pyriformis* group the rims are low or entirely undeveloped. In the *P. sulcatus* group, the rims are low in most species and in young specimens; in the typical species *P. sulcatus* and *P. halli*, they are 1 mm or more high. The flanges are 2 mm or more high in *P. spicatus* and *P. cherokeeus*, and the flange is high and serrate in *P. serratus* (Pls. 9, 10).

(7) *Extension of the deltoids.*—The extension of the deltoids above the mouth and spiracles is important and is characteristic of most species of the *P. sulcatus* group. In that group the deltoids reach the summit in the more primitive species and reach far above the summit in the most advanced species. The extended deltoids become flat radially and divide the upper ends of adjacent ambulacra. The extended tips are very frail and in most specimens of the *P. sulcatus* group they have been broken off in weathering and handling (probably not in the original depositing); the original specimen of *P. sulcatus* Roemer may have had the deltoid tips broken off, at least on the side toward the observer. In the other three groups of species the deltoids lack 1 or 2 mm of reaching as high as the spiracles.

In all species of *Pentremites* the deltoids extend between the upper ends of the ambulacra and divide the spiracles into two tubes, but the partition does not ordinarily show at the surface, except in *P. burlingtonensis*. In the anal tube the deltoid divides into two partitions, making three tubes in the anal spiracle (Pl. 13, fig. 5).

(*8*) *Flare of the deltoids.*—In *P. cherokeeus* and *P. halli* the upward extension of the deltoids becomes thin radially and flares outward over the interambulacral areas (Pl. 9, figs. 2–9; Pl. 13, figs. 4, 5). The flaring deltoids seem to be the main difference between these species and *P. sulcatus*, but a tendency to flare may be detected in many of the *P. sulcatus* group. The deltoids do not flare in forms with low or missing ambulacral rims, as *P. hambachi* and *P. gutschicki*. The flaring deltoids are very fragile, and the tips are broken off in most specimens of the *P. sulcatus* group.

(*9*) *Profile of the edges of the pelvis.*—The edge of the pelvis refers to the line from the base of an ambulacrum to the stem attachment. It is seen as a profile in a specimen viewed laterally, or in a figure of a lateral view. The sides of the pelvis are usually smooth, slightly concave lines, as in *P. godoni* and most members of the *P. sulcatus* group and other species with large pelvic angles. Most forms with the smaller pelvic angles, such as the *P. pyriformis* group, have straight-sided pelves. In a few forms, such as *P. chesterensis* and *P. broadheadi*, the basal plates are abnormally enlarged and extend below the lines of the radials. Such an enlargement of the basals is shown in Roemer's figure of *P. sulcatus* (Pl. 7, fig. 6).

(*10*) *Nodes on the basal plates and radials.*—The three basal plates, in a few species, have large, round, inflated areas, swellings, nodes, or bosses, first noted by Hambach (1903, p. 52) (Pl. 6, figs. 16, 17; Pl. 7, figs. 2, 3). They occur in the middle of the unpaired basal plate, and larger nodes occur on the paired plates centering at the ankylosed suture. They are outward swellings of the basal plates, rather than additions or callosities on the outside of the basal plates. They extend below the stem facet in most adult specimens, as seen typically in *P. tulipaformis* and *P. gutschicki*. They are less well marked on young specimens. There are low swellings in other species, as *P. godoni*, but they are scarcely sufficient to be called nodes. They are good specific characters for *P. tulipaformis* and *P. gutschicki*, but those species have other specific characters of equal importance. *P. sulcatus*, *P. broadheadi*, and *P. chesterensis* have swollen basal plates, even larger than nodes.

In one rare form, *P. nodosus* Hambach, there are nodes on the tips of the radials. Such nodes do not occur on any other species. The form without the nodes is known as *P. hambachi* Butts (Pl. 6, figs. 5–8). The presence of nodes is a poor character by which to distinguish species, but it seems best to recognize both *P. nodosus* and *P. hambachi* as separate species.

(*11*) *Shape of the sides of the vault and vault profile.*—The sides of the vault are usually moderately curved and make, with the summit, a parabolic curve, as in *P. godoni*, *P. pyriformis*, *P. welleri*, and *P. biconvexus*. In some specimens the sides are strongly curved, where the lateral profile, including the summit, is nearly circular, as in *P. sulcatus*, *P. gutschicki*, *P. obesus*, *P. conoideus amplus*, *P. godoni pinguis*, and *P. hambachi*. The sides may be only slightly curved but not quite straight, as in *P. symmetricus*, *P. pyramidatus*, *P. abruptus*, *P. malotti*, *P. cervinus*, *P. gracilens*, *P. conoideus*, and *P. rusticus*, where the lateral profile of the vault is that of a truncate pyramid. In *P. rusticus* the sides of the vault are only slightly curved and tend to be parallel, giving the specimen a barrel shape.

(*12*) *Position of greatest width.*—Ordinarily the greatest width of a specimen is measured from lower tip of one ambulacrum to the ambulacrum 144° from it. In very convex or gloublar specimens, the greatest width is above the ambulacral tips, e.g., *P. burlingtonensis*, *P. godoni pinguis*, and *P. hambachi*. In forms with very long pelves, such as *P. okawensis* and *P. clavatus*, the greatest width is supramedian. It is a function of the pelvic angle and the vault-pelvis ratio.

(*13*) *General shape of the specimen.*—The general form of a pentremite can usually be given in one word; it is a composite of all structural characters considered at once. All the pentremites, are bud-shaped, as the name Blastoidea indicates. Say points out (1820a, p. 35) that "Persons who have not devoted their attention to the affinities of natural objects, have regarded it as a petrified *nut* or *Althea bud.*" *P. godoni* does have much the shape of Hollyhock buds (*Althaea rosea*), which also have five segments, a base and a stem, but there the similarity ends. The general shape fairly well places a specimen in one of the four groups into which it is convenient and natural to divide the pentremites. In the *P. conoideus* group specimens are ovoid to conoidal or pyramidal; in the *P. godoni* group they are subglobular to conoidal and budlike; in the *P. pyriformis* group most specimens are pear-shaped, with long pelves; in the *P. sulcatus* group specimens are broad and budlike. *P. halli* and *P. malotti* are pentagonal in side view as well as in basal view. A few species can be referred to by particular words, indicating shape: *conoideus, obesus, ovoides, clavatus, platybasis, abbreviatus, pyriformis, pyramidatus,* and *elongatus*. The general shape of the specimens, together with the depth of the ambulacra, make it possible to recognize many species without measurements.

(*14*) *Size of the specimen.*—Adult individuals of each species of *Pentremites* have their normal size, and adults have different proportions from young specimens of the same species. *P. platybasis* is an example of a small form, normally about 10 mm long. *P. pyriformis*, 15 to 20 mm long, is an example of a species of medium size. *P. welleri* is an example of a form of large size, 25 to 30 mm in length, and *P. obesus* and *P. maccalliei* are examples of very large forms, up to 62 mm. Only adults should be used in measurements and descriptions. In practice, it may be assumed that the larger specimens, for example in a group of five or more, are adults. Nepionic specimens of pentremites are almost never found; early neanic and gerontic specimens are rare. Young and normal ephebic stages are the most common of ontogenetic stages in collections. The size of the specimen has some bearing on its specific identity, but care should be used in considering the ontogenetic stage.

## SECONDARY CHARACTERS

(*15*) *Width of ambulacra.*—The ambulacra in *P. conoideus* and its varieties are narrow, about 2 mm wide at the midpoint, and the sides are almost parallel. In most species of *Pentremites* the ambulacra flare slightly, and the upper ends are from 3 to 5 mm wide. In the *P. sulcatus* group the ambulacra are wide and deep, and the sides tend to be parallel; the width is 6 to 8 mm reaching the greatest width, 8 to 10 mm, in the largest specimens of *P. sulcatus*, *P. obesus*, and *P. maccalliei*. The width of the ambulacra varies also with the ontogenetic stage of the specimen. Although the width of the ambulacra at some point is a natural fact for each specimen, the widths of the ambulacra are not reliable specific characters, even if the measurements are reduced to some arbitrary standard, as was done by Clark (1917, p. 363–367).

(*16*) *Length of the ambulacra.*—The length of the ambulacrum is measured from tip to top, along the curvature, or by a chord from tip to top, rather than by the projection of the length on the axis of the calyx. It is used in calculating the ambulacral ratio but usually omitted in specific descriptions. The length varies with the specimen and with the ontogenetic stage and is of little value in identification. In most specimens it is less than the height of the vault, but its importance in classification is submerged in the vault-pelvis ratio. The ambulacral ratio, length divided by width, is of some value in identification. Length divided by width varies from a quotient of 1 in the young to 4 or 5 in the adult of the same specimen. The close relatives of *P. conoideus* have an ambulacral ratio of 4 to 8; for most other pentremites the ratio is about 3 to 4. The wider the summit of the calyx the wider the upper end of the ambulacra, since the five ambulacra touch at their upper ends and make a circle regardless of the length of the ambulacra. *P. cherokeeus* has unusually narrow ambulacra, 4–6 mm, and a ratio of 5 or 6.

(*17*) *Number of lateral grooves in 3 mm.*—The number of lateral grooves on one side of an ambulacrum varies with the youth or maturity of the specimen: it is higher in young specimens, about 12 in 3 mm, and in adults is generally 8 or 9 in 3 mm. The grooves are also closer together in the upper end of the ambulacra, about 12 in 3 mm, whereas there are about 9 in 3 mm in the lower and middle parts of the ambulacra. *P. altus* has the fewest, 6 or 7 in 3 mm; *P. symmetricus* has 7 in 3 mm; some of the *P. sulcatus* group have 9 or 10 in 3 mm. The lateral grooves of the giant specimens, such as *P. obesus*, *P. sulcatus*, and *P. maccalliei* are about the same as in the smaller adult, *P. godoni*, 8 or 9 in 3 mm. Ulrich stated (1917, p. 262) that *P. lyoni* and *P. welleri* ". . . are constantly distinguished by the narrower and therefore more numerous poral pieces" in *P. lyoni*. The size of the grooves is not a reliable character for specific differentiation, but the number of grooves in 3 mm varies with the species as well as with the ontogenetic age in the specimen and the position on the ambulacrum.

(*18*) *Slope of lateral grooves.*—The transverse ridges, between which the lateral grooves occur, alternate on the two sides of the median groove. They slope upward and meet at angles of 90° to 120° in young specimens, becoming more nearly transverse with ontogenetic age. Species with the narrowest ambulacra, as *P. conoideus* and *P. altus*, have more sloping ridges and grooves, especially in the young, than do forms with wide and concave ambulacra, such as the *P. sulcatus* group.

(*19*) *Fine structures of the ambulacra.*—Some of the detailed structures of the ambulacra have been well illustrated by Hambach (1880, Pl. A, fig. 15; 1903, Pl. 2, figs. 2, 5) and by Etheridge and Carpenter (1886, Pl. 1, fig. 2). The finer structures are: (a) minute ridges and grooves which are transverse on the main ambulacral ridges and swing around their ends; (b) minute ridges and grooves on and parallel to the main ridges; (c) inclination of the outer ends of the ridges; (d) grooves at the outer ends of the transverse ridges; (e) round pits at the downward flexure of the transverse ridges; (f) pits at the suture between side plates and lancet plate (rare). The fine structures are too small to show well on photographs.

(a) The minute ridges and grooves are well shown by Etheridge and Carpenter (1886, Pl. 1, fig. 2), whose figure is partly reproduced in Figure 3 of Plate 13. The minute ridges are larger on the sides of each main ridge and become smaller as they swing around the inner ends of the ridges. The number of toelike ridges varies from five, in *P. conoideus* (Pl. 1, fig. 1), to seven or eight in *P. symmetricus* (Pl. 1, fig. 5).

(b) On the best-preserved ambulacra there are very fine longitudinal ridges and grooves on the main ridges and grooves. The minute ridges are fewer, two or three, on sharp and narrow ridges, as on *P. symmetricus* (Pl. 1, fig. 4) and *P. tulipaformis* (Pl. 1, fig. 11). There are four to six minute ridges on wider ridges, as on another, more typical specimen of *P. symmetricus* (Pl. 1, fig. 5).

(c) The outer ends of the transverse ridges turn obliquely downward in all species of *Pentremites*. The oblique band is widest in the forms with convex or flat ambulacra, as *P. conoideus* (Pl. 1, fig. 1), *P. godoni* (Pl. 1, fig. 2), and *P. pyramidatus* (Pl. 1, fig. 6), although it may be narrow and not well marked, as in *P. godoni angustus* (Pl. 1, fig. 3) and *P. symmetricus* (Pl. 1, fig. 5). In forms with concave ambulacra, the *P. sulcatus* group, the transverse ridges are more nearly parallel on the two sides of the ambulacral groove, and the oblique outer bands of the transverse ridges are narrower and not so oblique (Pl. 1, figs. 9–12). In *P. macalliei* (Pl. 8, fig. 3), and in others of the largest forms, the oblique bands are narrow and not well marked (Pl. 8, figs. 4, 5; Pl. 9, figs. 9–11).

(d) The transverse ridges have short grooves on the outer ends; in most specimens the grooves are confined to the band of oblique ends of the ridges (Pl. 1, figs. 1, 3, 5, 12). In some specimens the secondary grooves extend from the lateral pores inward past the junction of the side plates and the lancet plate (Pl. 1, figs. 2, 4, 6, 7, 9). In some specimens the secondary grooves extend almost the full length of the transverse ridge (Pl. 1, figs. 10, 11; Pl. 13, fig. 6). There are good specimens of *P. fohsi* which agree in every other respect with the type (Pl. 13, fig. 6) but do not have secondary grooves on the transverse ridges.

(e) Some well-preserved specimens have round pits at the outer ends of the transverse ridges at the downward bend. The pits may have been for the attachment of the pinnules (Pl. 1, figs. 1, 3, 5, 6, 7, 8, 9–11).

(f) In rare cases there are oval pits at the junction of the side plates and the lancet plate, as in *P. halli* (Pl. 13, fig. 4a).

Perhaps there are smaller features which have not been observed. Few species, if any, can be identified by the microscopic features of the ambulacra alone, but the minute features of each group of closely related species seem to be essentially the same. Most species are identifiable without the minute structures seen on the best-preserved ambulacra, but the concavity, flatness, and convexity of the ambulacra must always be considered.

Ulrich remarked (*in* Ulrich and Smith, 1905, p. 64),

"Much of the uncertainty prevailing among paleontologists endeavoring to identify species of *Pentremites* is due to the fact that in describing and classifying the species authors have paid more attention to the comparatively unimportant variations in the general form of the theca than to the more constant structural features displayed on the surface of the ambulacral areas."

Different groups of species have finer characteristics of the ambulacra, and in some excellently preserved specimens the form can be identified with a small group of species. The ambulacrum of the specimen of *P. symmetricus* (Pl. 1, fig. 4; Pl. 4, fig. 13) is much like that of *P. pyramidatus* (Pl. 1, fig. 6) and that of *P. kirki* (Pl. 1, fig. 7), yet the specimen is identified as *P. symmetricus* on the basis of the more constant features, length over width, vault over pelvis, and pelvic angle.

(*20*) *Angle of the vault.*—In species where the vault sides are very little curved, such as *P. symmetricus*, *P. altus*, *P. maccaliei*, and *P. godoni angustus*, it is possible to measure with a protractor the general angle of convergence of the vault. In those species it is 30° or 40°. In forms with moderately to strongly convex vault sides, such as *P. godoni*, *P. obesus*, *P. okawensis*, and *P. sulcatus*, it is not possible to state the convergence of the sides of the vault, since there is no constant angle to two tangents to opposite curves. The angle of the vault is therefore useful only where the sides of the vault are nearly straight, and even there it is of negligible value.

(*21*) *Angle between pelvis and vault.*—The pelvis-vault angle is not a reliable feature for distinguishing species, since it is usually the angle between the concave sides of the pelvis and the convex sides of the vault. It may be as low as 90°, as in *P. godoni abbreviatus*, or as high as 150°, as in *P. clavatus*. The more usual angle is 110° to 120°. Essie Smith (1906, p. 1222–1226) measured the angle between the ambulacral area and the base for hundreds of specimens of *P. conoideus* and found the angle to vary from 98° to 125°, averaging between 110° and 115°.

(22) *Shape of the summit.*—The summit is the region about the mouth, including the spiracles and the upper ends of the ambulacra. In the *P. sulcatus* group, in which most of the deltoids extend above the mouth or summit, the extended deltoids constitute the apex of the calyx. In most specimens the summit is slightly convex, as in *P. godoni* and *P. conoideus*; in some it is nearly flat, as in *P. symmetricus*; in most of the *P. sulcatus* group it may be considered to be concave. The summit should be noted in both group and specific identification. The shape of the summit is determined largely by the extension of the deltoid plates.

(*23*) *Width of the summit.*—The summit is generally two-fifths to two-thirds the width of the calyx. In some species it is unusually narrow, as in *P. godoni angustus*, *P. gutschicki*, and *P. conoideus*, and in others very wide, as in *P. rusticus* and *P. godoni major*. The width of the summit is a fact to be noted in specific identifications, but it is not possible to measure it exactly, and it varies greatly in the same species; it may be omitted in specific descriptions, except for the very narrow and the very wide summits. The width of the summit affects the angle of the vault; the wider the summit the smaller the angle of the vault.

(*24*) *Length of the deltoid plates.*—The deltoid plates are ordinarily from three-tenths to four-tenths the length of the ambulacra. The only known species in which it is over half of the length of the ambulacra is *P. pediculatus*, a young specimen of *P. welleri*. The deltoids grow by addition at their bases; the tips of the radials grow by additions at their upper ends and along the vertical suture. In many specimens it is difficult or impossible to distinguish the real deltoid suture from the striations near the suture. See Ulrich (1917, Pl. 2, figs. 34, 35), in which the deltoid suture in the photograph, Figure 34, is at least 2 mm higher than in the retouched figure of the same specimen, Figure 35. The importance of the long deltoids of *P. pediculatus* (Ulrich, 1917, Pl. 2, fig. 44) may therefore be doubted, and the relative length is of negligible importance.

(25) *Ontogenetic stages.*—Young specimens have longer pelves; therefore the pelvic angle is smaller than in the adult, the ambulacra are proportionally shorter, with fewer transverse grooves which are oblique, and the deltoids are very short or even missing in neanic specimens. *P. praematurus* is probably a young specimen of some other species, as is true of *P. benedicti* Rowley, the nepionic stage of *P. conoideus*. Hambach (1903, Pl. 6) figured an ontogenetic series of *P. sulcatus*, all specimens from the same locality and age (Pl. 12, figs. 1–12). In that series, Figures 1–4 are like *P. angularis* Lyon, Figures 5–7 are like *P. halli* new species, Figure 8 is like *P. fohsi* Ulrich, Figures 9–11 are like *P. sulcatus* (Roemer), and Figure 12 is like *P. spicatus* Ulrich. Hambach's own species, Figure 9 of Plate 4 in the same publication, *P. serratus* Hambach, may be a gerontic specimen of the same series (Pl. 10, fig. 7).

Essie A. Smith (1906, p. 1219) studied the ontogeny of *P. conoideus*, (Pl. 11, figs. 1–19) and showed that the earliest stage, the nepionic, is less than 1 mm long and that the pelvis consists of three basal plates, as in the adult, and five radials; no ambulacra or deltoid plates were preserved (Pl. 2, fig. 11). The pelvic angle is about 40°. This is the stage named *P. benedicti* by Rowley (1900) (Pl. 2, fig. 15). Miss Smith therefore concluded that the ancestral form was a *Codaster*, since the early stage has a small pelvic angle like that of *Codaster*. *Codaster* may be a distant ancestor, but the immediate ancestor of *Pentremites* was probably the genus *Pentremitidea*. Young stages of *P. conoideus*, in which the pelvic angle is about 80° and the vault-pelvis ratio about 1, were identified as a separate species, *P. koninckanus*, by Hall (1858, p. 656) (Pl. 2, figs. 13, 14). The writers agree with Whitfield (1882, p. 44) and Miss Smith that *P. koninckanus* is only the neanic stage of *P. conoideus*. Young specimens have been identified by Ulrich (1917, Pl. 2) as *P. praematurus* and *P. buttsi*. It is very difficult to identify young specimens without a series from neanic to ephibic specimens. Young specimens of the *P. conoideus* group are not specifically distinguishable, as are young specimens of the *P. godoni*, *P. pyriformis*, and *P. sulcatus* groups.

## NEGLIGIBLE CHARACTERS

(26) *Supposed dwarfing.*—Miss Smith concluded (1906, p. 1237) that the small specimens of *P. conoideus* in the Old Cleveland quarry 1 mile north of Harrodsburg, Indiana, had been dwarfed. Evidence for dwarfing is inconclusive. The small pentremites were probably carried by water currents to the place of deposition, with small corals, brachiopods, small adult specimens of ostracodes, gastropods, pentremites, *Endothyra*, spicules, and broken pieces, the same size as supposed dwarfs, of bryozoans, brachiopods, crinoid stems and plates, and other debris.

(27) *Total number of transverse grooves.*—The number of lateral grooves in an ambulacrum varies with the length of the ambulacrum and is therefore a function of

both the ontogenetic age of a specimen and the size of the adult. In *P. platybasis*, the number is 20 to 25; in *P. godoni*, it is from 40 to 50; in *P. maccalliei*, it is as high as 104. The total number is not a specific character. Lyon (1860, p. 633) states that the number for *P. elegans* is 63. Few later students have mentioned the number of lateral grooves for a species. The total number will be about 3 times the total length of the ambulacrum in millimeters. The grooves incline upward and inward at the lower small end of the ambulacrum and are transverse at the upper end.

(*28*) *Shape and size of the lateral pores.*—The lateral pores are elongate and not well exposed in well-preserved specimens. They are best seen in slightly weathered specimens and are then oval to round and about a sixth of a millimeter in diameter. Their character is not considered useful in specific identification (Pl. 1, figs. 6–9).

(*29*) *Width of lancet and side plates.*—The lancet plate is the long lance-shaped plate, with alternating transverse grooves, and a median groove, (Pl. 1, figs. 10, 11). The side plates border the lancet plates and bear smaller grooves and the lateral pores. The suture between the lancet plate and the side plates inclines inward and toward the median groove. The side plate is narrower in some well-preserved specimens but about as wide as one side of the lancet plate; in weathered specimens it is usually less than the width of half of the lancet plate. The relative width of the lancet and side plates is of little or no specific value.

(*30*) *Median groove of the ambulacrum.*—The median groove extends from the bottom to the top of the ambulacrum, divides it into two equal parts, and extends into the mouth. It is narrow and shallow, but conspicuous in forms with convex ambulacra, such as *P. conoideus* and its allies. It is about half a millimeter wide in *P. godoni* and its close allies and about half a millimeter wide and deep in *P. sulcatus* and its allies. It is least conspicuous in forms with U-shaped ambulacra, such as *P. tulipaformis* and *P. gutschicki*. It does not appear to be a specific character. The median groove is not a suture.

(*31*) *Shape and size of the mouth.*—The mouth is a five-pointed star, with the median grooves of the ambulacra entering at the points. The shape seems to be the same for all species, and the size is larger in the larger specimens. It is about 1 mm in diameter in *P. godoni* and about 2 mm in diameter in *P. sulcatus*; in young specimens the anal spiracle crowds the mouth and is crescentic. The writers have not seen plates in the mouth or the spiracles, as figured by Shumard (1858, Pl. 9, fig. 4) (Pl. 2, fig. 10.), but grains of sand in the mouth and spiracles have much the same appearance. Dr. Courtney Werner, of Washington University, states that he has a specimen which has plates over the mouth.

(*32*) *Shape and size of the spiracles.*—Four of the spiracles are of equal size and in well-preserved specimens are ovoid with a point toward the mouth. Each spiracle is divided at a depth of about 1 mm by the thin plate of the deltoid extending inward. The fifth opening, the anal, is about $1\frac{1}{2}$ times the size of the other four spiracles but of the same shape and is divided into three tubes by the division of the adjacent deltoid into two parts. The spiracles are not covered by plates, even in the best-preserved specimens. Such covering plates may have been grains of sand or pieces of pinules in the mouth and spiracles, as Hambach insisted (1903, p. 15, 16). One specimen in the U. S. National Museum has a conical structure covering the mouth, composed of small rods. Another has a structure, similar to the anal tube of crinoids, which covers both the mouth and spiracles. Since such structures are almost never found, they are of no value in specific identifications (Hambach, 1884a, p. 541, Fig. 2; 1903, p. 14, Figs. 6, 7).

(*33*) *Basal suture of the deltoids.*—The suture, where the deltoid joins the upper ends of two contiguous halves of the radial plates, is V-shaped, with an angle of 60°–90°. The sides of the V are usually straight but they are sigmoid curves in *P. obesus* (Pl. 8, fig. 4), *P. maccalliei*, *P. fohsi*, *P. robustus*, and other species. In forms of moderate size the suture is normally a simple V, but a double sigmoid suture also occurs in the same species with the V, as in *P. godoni* and *P. pyriformis*. The suture is the place at which growth takes place on all three of the plates involved. The growth may be somewhat irregular both in angle and shape of the suture. The deltoid-radial suture is not a reliable criterion even for varietal identification. Growth lines parallel the suture both above and below (Pl. 13, fig. 1; Pl. 8, figs. 3, 4).

(*34*) *Shape of the basal plates.*—There are three basal plates, as is true of many crinoids, two of which were formed by the ankylosis of two pairs of plates; the third is unpaired. The unpaired plate is quadrangular and about half the size of the paired plates or a little larger. The unpaired plate is nearly always right anterior (Figure 5 of Plate 13 is an exception). The pattern formed by the three basal plates is pentagonal, generally with nearly straight sides; in some specimens it is substellate, as in *P. rusticus* (Pl. 3, fig. 19), *P. godoni*, and *P. conoideus*. There is no consistent difference in different species or in different groups of species. The downward protuberance of the basal plates in *P. sulcatus*, *P. chesterensis*, and *P. broadheadi* are constant characters (Pl. 7, figs. 5–7; Pl. 8, fig. 1; Pl. 9, fig. 1.).

Many species with nearly flat base, such as *P. godoni*, *P. platybasis*, and *P. tulipaformis*, have a triangular depression about the stem cicatrix. Inasmuch as the sutures between the three ordinary basal plates extend through the corners of the small triangle to the center of the stem, workers since Lyon have not recognized the three small basal plates of Lyon (1857, p. 468); Hambach (1880, p. 146) rejected the three extra basal plates, as do the present writers (Pl. 6, fig. 16b).

(*35*) *Radial pelvic ridges.*—Ridges extend from the lower tips of the ambulacra toward the stem, are sharper near the ambulacra, and usually fade out before reaching the basal plates (Pl. 3, figs. 9b, 29c). The ridges are sharpest and longest in species with concave interambulacra, such as *P. springeri* and *P. welleri*. There is no particular radial ridge on the base of the *P. sulcatus* group, except at the points of the ambulacra.

(*36*) *Length of the stem.*—The length of the stem of the pelvis depends upon whether or not a new columnal button has been cut off, not upon the species. The columnals were formed at the center of the basal plates and in some specimens were not entirely cut off in formation. *P. pediculatus* is some other species, probably *P. arctibrachiatus* or *P. welleri*, with a stem ready to form another columnal (Pl. 4, fig. 31).

(*37*) *Diameter of the stem.*—One species, *P. gemmiformis*, was founded because it had an abnormally large stem for a small pentremite. It is probably not a valid species. The stem is 1 mm in diameter in very small specimens, about 2 mm in normal specimens of about 20 mm length, up to 3 or 4 mm in diameter in *P. sulcatus*, and up to 5 mm in diameter in *P. obesus*, counting the columnals still attached. The stem size is a function of the ontogenetic stage and the size of the specimen. It may also be hypertrophied. It is not of specific value (Pl. 5, figs. 5, 6).

(*38*) *Striations on the plates.*—The plates grow by addition of parallel increments on both sides of the sutures between similar plates, between basals and radials, and between radials and deltoids (Pl. 13, fig. 1). The growth lines are most conspicuous on the radial plates adjacent to the suture. Such ornamentation is a matter of preservation or development due to weathering and has nothing to do with specific identities. One species, *P. laminatus* Easton (1943, p. 136), was named for the laminations of the growth lines.

(*39*) *Abnormalities.*—Several species have been erected on the basis of abnormalities of the plates, and such forms are probably not real species. *P. broadheadi* may be only an abnormal *P. obesus*; *P. spinosus* may be only an abnormal *P. sulcatus* or even *P. halli*; *P. nodosus* seems to be an abnormal specimen of *P. hambachi*. *P. nodosus* of Weller (1920, p. 356, Pl. 4, fig. 25) is very close to *P. hambachi*.

Four-sided specimens, one six-sided specimen, and specimens with other abnormalities have been described and figured by Hambach (1903, p. 3; Pl. 3, figs. 7–16) and by Etheridge and Carpenter (1886, p. 39; Pl. 2, figs. 8–12). There are four-sided and six-sided specimens of *P. conoideus* in the Dr. C. A. Malott collection of *Pentremites*. They are neither different species nor varieties but abnormal specimens of their species.

(*40*) *Summit plates and ambulacral covering plates.*—Hambach (1884a, p. 541) described a tubular pyramid covering the mouth and spiracles. *P. fohsi* in the U. S. National Museum has this structure, and Prof. Courtney Werner also has such a specimen. Shumard (1858, p. 243, Pl. 9, fig. 4) and Billings (1869, p. 69) figured little plates covering the mouth and spiracles. There are specimens of *P. conoideus* with the mouth and spiracles filled with grains of rock, fragments of fossils, and oölitic grains, but the writers have seen no specimens with definite covering plates. There may also have been plates over the ambulacral groove (Etheridge and Carpenter, 1886, p. 163; Pl. 1, fig. 8), but there are no specimens exhibiting such plates in the collections studied. Such structures are of no specific value, since they are rarely preserved, and their significance is in doubt. Etheridge and Carpenter (1886, p. 1–128) discussed extensively the morphology and possible function of structures of *Pentremites* and other blastoids, and little information has been added in the past 70 years.

(*41*) *Pinnules.*—Tenuous jointed arms attached to the edges of the ambulacra (Pl. 13, fig. 2) were discovered by Roemer (1848) and have been described or figured by Billings (1870, p. 228), Hambach (1880, p. 151), Etheridge and Carpenter (1886, p. 63; Pl. 11, figs. 16–17), and Butts (1926, Pl. 59, fig. 4). Their rarity precludes their use in systematic descriptions. In rocks replete with *Pentremites* and small fossils, which should include pinnules, there are no fragments of pinnules. There is a specimen at the U. S. National Museum which does show them attached. Recently pinnules have been called brachioles.

(*42*) *Hydrospires.*—Under the side plates there are 10 groups of hydrospires into which the lateral pores opened. Each group consists of four to nine hydrospires. The rarity of their preservation and the difficulty of examining them precludes their use in specific identification (Pl. 7, fig. 9).

### NONSTRUCTURAL FEATURES

(43) *Geographic occurrence.*—The geographic range of species of invertebrates is not world-wide. In general, one may expect a different species at a distance of about 1000 miles from the locality of the type specimen. Geographic species and varieties are of common occurrence, both in recent species of invertebrates and in fossil invertebrates. But it should not be assumed that specimens found in Montana are *ipso facto* different from similar species found in Kentucky. Clark (1917, p. 363–367) described three new species from Montana which seem to differ in no constant way from species already described from Kentucky. *P. cherokeeus* described by Troost from east Tennessee seems to be exactly the same form found in Indiana from the Glen Dean or equivalent horizon.

(44) *Stratigraphic occurrence.*—A species of marine invertebrates has a usual range of about a stratigraphic series (one-third of a system) or an epoch. The pentremites as well as the crinoids seem to have a range of about half of a series. Ulrich considered the pentremites guide fossils, species characteristic of given horizons (1917, p. 143, 169). Few if any of the species of *Pentremites* he identified and described ranged from one formation to another. Most of the species Weller (1920, p. 314, *et seq.*) identified and described were restricted to one formation, but he thought that *P. princetonensis* ranged from the St. Louis to the lower Chester. Formations can be identified reliably by determining the association of pentremites in different formations. Species range from one group of formations into another, about two-thirds of a series (Figs. 1–5).

(45) *Abundance.*—Abundance of specimens of a species is of importance in identifying a formation. *P. conoideus* ranges from the middle Harrodsburg into the St. Louis but is abundant only in the top of the Harrodsburg and base of the Salem limestone. *P. godoni* ranges from the Renault into the Golconda but is found most abundantly in the Beaver Bend and Paint Creek formations. *P. obesus* seems to be confined to the Golconda formation. *P. sulcatus* and its varieties occur abundantly in the Glen Dean, but some of them extend into the Menard limestone of the upper Chester. *P. tulipaformis* is abundant in the Golconda, the Glen Dean, and the Kinkaid limestone. *P. gutschicki* is abundant in the Kinkaid limestone.

### GEOLOGIC HISTORY OF *PENTREMITES*

Very few specimens of the genus *Pentremites* appear in the Lower Mississippian Burlington limestone, and there are none in the Kinderhookian. *Pentremites* first became abundant in the late Harrodsburg and Salem limestones. There are a few species in the St. Louis and Ste. Genevieve limestones. In the lower Chester there are 11 species in the Renault and Beaver Bend formations and 19 species in the Paint Creek formation.

*Pentremites* are abundant in the middle Chester Golconda limestone, from which about 30 species have been reported. Ulrich (1917, p. 226) lists 22 species from the

Glen Dean formation of Kentucky. In the upper middle Chester, Glen Dean formation, with 34 species, the genus reaches its acme in number of individuals, number of species, and size of specimens. The genus decreases suddenly in the upper Chester, in which about 12 species occur. Large specimens of the *P. sulcatus* group and smaller specimens of the *P. pyriformis* group have been found in the Menard, Clore, and Kinkaid limestones.

Outside of Arkansas, the genus becomes extinct in the upper Chester. In Arkansas and Oklahoma, Mather (1915, p. 100) found *P. angustus* and *P. rusticus* in the Brentwood limestone, correlated then and now with the Lower Pennsylvanian. Morgan (1924, p. 198) found the same species in the Wapanucka formation, Lower Pennsylvanian, of southern Oklahoma. This is the last of the pentremites. The age of the beds can still be questioned; it may be Chester—according to the pentremites, lower and middle Chester. Warren (1927, p. 48) found large specimens of *Pentremites* in the Rundle limestone of the Banff Area, Alberta, which he considers (p. 34) to be Mississippian in age.

The earliest species of *Pentremites*, of the Lower Mississippian, have very short pelves, narrow convex ambulacra, concave interambulacra, no ambulacral rim, and short deltoids (the *P. conoideus* group). The specimens are melon-shaped, greatest diameter median, with ends slightly convex and much alike. The same general characters obtain up to the Warsaw and Salem, where the form becomes conoidal by the widening of the base. The earliest *Pentremites* are similar to the Devonian species of the genus *Pentremitidea*, differing only in lacking the lancet plates covered by the side plates.

In the Ste. Genevieve the form becomes budlike, with the base a little more elongate and the ambulacra wider and less convex; the ambulacral rim appears but is low, the interambulacra becomes less concave, the deltoids stay below the summit, and the size of specimens does not increase noticeably. The budlike form reaches its best development in the lower Chester (*P. godoni* group) and continues sparingly into the Glen Dean, perhaps to the Kinkaid.

In the lower Chester the shape becomes in part pyriform (the *P. pyriformis* group) the pelvis becomes nearly as long as the vault, and in the middle Chester most species have long pelves.

In the middle Chester the general bud shape of the calyx is retained in the *P. sulcatus* group, but there is a great accentuation of the depth of the ambulacra, the height of the ambulacral rim, the length of the deltoid, and the size of the calyx.

The *P. pyriformis* and *P. sulcatus* groups are characteristic of the middle and upper Chester.

The ambulacra in the earliest forms, Lower and Middle Mississippian, are convex and narrow, with the side plates almost invisible. The ambulacra become wider and nearly flat in the Ste. Genevieve. They are slightly concave in part in the lower Chester, more concave in the middle Chester, and culminate in the deep U of the *P. sulcatus* group of the Glen Dean and upper Chester.

The interambulacral area is concave in the earliest forms, the *P. conoideus* group, becomes less concave in the *P. godoni* and *P. pyriformis* groups, but is deep again in the *P. sulcatus* group.

The ambulacral rim, the outward extension of the rims on the radials and deltoids, is missing in forms before the Ste. Genevieve, where it is first noticeable. The rim appears in the *P. godoni* group, is scarcely detectable in the *P. pyriformis* groups, but becomes a high flange in most species of the *P. sulcatus* group.

The deltoid plates are short, *i.e.*, they terminate below the spiracles in all groups but the *P. sulcatus* group. In this group they are even with the spiracles in the more primitive species and in young stages. The deltoids grow outward radially, or flare, because of the development of the ambulacral rim, as in *P. halli* and *P. cherokeeus*, and are thin radial plates, very fragile and easily broken off in weathering and handling of specimens. The deltoids grow upward and outward as a result of development of the ambulacral rims and may extend upward as much as 5 mm above the spiracles, as in *P. spicatus*. Some of the writers' specimens of *P. sulcatus* have deltoids extending 1 or 2 mm above the spiracles. They may have been partly broken off in Roemer's figured type specimen (Pl. 7, fig. 6).

The species of *Pentremites* from the Brentwood and Wapanucka limestones, supposedly of Lower Pennsylvanian age, belong to the *P. godoni* group and are like lower Chester forms.

It is remarkable, as noted by Etheridge and Carpenter (1886, p. 134), that only one species had been found in the Keokuk limestone. Keyes (1894, p. 134) records *P. conoideus* from the Keokuk limestone of Booneville, Missouri.

The geologic occurrence of species and a comparison of the development of structures in time, *i.e.*, morphogenesis, are fairly reliable and satisfactory criteria for determining the phylogeny of the genus, and for determining stratigraphic horizons.

Figure 1 shows the geologic occurrence of the groups of *Pentremites*, the abundance in species (indicated by width of lines), and the phylogeny of each group, with the time it arose.

## ONTOGENETIC DEVELOPMENT OF *PENTREMITES*

The embryology of the *Pentremites* is unknown and probably undeterminable, because there were no hard parts in the embryonic stage. But Twenhofel and Shrock's statement (1935, p. 171) that "Nothing is known of the ontogeny of the Blastoidea" is too sweeping, for much is known of the epembryonic ontogenetic development of *Pentremites*, particularly of *P. conoideus* and *P. sulcatus*.

Hambach (1884a, p. 542, Fig. 3) indicated that the plates of *P. sulcatus* grow by lateral or peripheral additions. Hambach (1903, p. 5, 8) also described the development of the deltoid plates by lateral expansion from terminal plates in the young to normal lateral plates in adult *Pentremites*. He also notes (1903, p. 8) the similarity of young *Pentremites* to adult *Pentremitidea*; he thought the two genera were the same: "Otherwise the young specimens of *Pentremites sulcatus, pyriformis*, etc., would belong to the genus *Pentremitidea* during their juvenile state, and afterwards when fully developed to *Pentremites*." On the basis of the recapitulation of the *Pentremitidea* stage, that genus is undoubtedly the ancestor of *Pentremites*, as borne out by comparison of adult structures and by geologic occurrence.

The ontogeny of *P. conoideus* is the best known of any species of the genus. Essie A. Smith (1906, p. 1232) studied and illustrated an ontogenetic series of *P. conoideus*

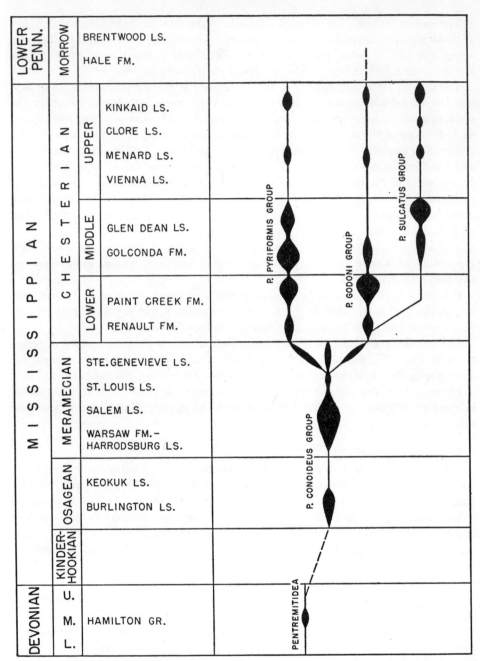

FIGURE 1.—*Geologic occurrence and phylogeny of the groups of* Pentremites

from a length of 0.82 mm up to 22 mm long. The ontogenetic series studied by Miss Smith, with additional specimens, is refigured on Plate 11.

The youngest specimens, in the ananepionic stage, about 1 mm in length, are conical and nearly circular in cross section and have a basal angle of 40°. There are three basal plates and five radial plates; the deltoids are not yet formed. According to Hambach (1903, p. 5) the median sutures of two larger basal plates are not ankylosed, but this was denied by Miss Smith (1906, p. 1232). The ambulacra, mouth, and spiracles are missing, and the calyx is little more than an empty conical cup, with indications of the ambulacral sinuses in the youngest stage (Pl. 2, fig. 11; Pl. 11, fig. 1).

At a length of 2 mm the basal angle is still 40°–50°, the mouth and spiracles are surrounded by bony parts, the ambulacra are closed, but no median groove, lateral pores, or poral pieces are discernible; a columnal is ready to be cut off from the basals. The deltoids are not yet developed. The summit is only slightly convex. It is the metanepionic stage (Pl. 2, fig. 12; Pl. 11, fig. 2).

At 2½ mm length the basal angle is still less than 50°, the ambulacra have four or five pores on a side and a median groove, four or five converging poral pieces in a side, and the side plates are not discernible; the deltoids appear as vertical plates between the ambulacra but do not appear on the sides of the cup. This stage is the "*P. benedicti*" Rowley, or paranepionic stage (Pl. 11, fig. 3).

At 5 mm the vault and pelvis are of equal length, the basal angle is 60°, the radials have grown much more than the basals, the vault is bulbous, side plates appear, but the deltoids are still confined to the summit. This is the neanic stage or "*P. konincka-nus*" Hall. It persists to 7 mm in length (Pl. 11, fig. 9).

Young ephebic specimens are from 8 to 10 mm in length. The basal angle expands from 100° to 120°, and the deltoids appear at the top of the interambulacral area. Normal adults begin at about 10 mm; the normal basal angle is 135° but may expand to 150°.

During ontogenetic development, the basal angle expands from 40° to 150°; the ambulacra increase in length by addition to the radial and deltoid plates and by addition of poral pieces to the top of the earlier poral pieces. The early poral pieces meet at the median groove forming an inverted V, and the later ones meet in a straight line. The last poral pieces added are narrower than the previous ones, even narrower than the ones at the lower tips of the ambulacra. The side pieces emerge in the neanic or "*P. koninckanus*" stage. The interambulacra are at first flat but become more and more concave during ontogeny.

The plates grow by small additions on both sides of the sutures, not by enlargement throughout, as bones grow (Pl. 13, fig. 1). The additions parallel to the sutures make the fine lines which have frequently been described as mere ornamentation. In the growth of the calyx the vault grows more than the pelvis in the *P. conoideus*, *P. godoni*, and *P. sulcatus* groups. In the *P. pyriformis* group only, the growth of the pelvis keeps pace with or exceeds the growth of the vault; in this group the shape, especially the pelvic angle, does not change greatly from the nepionic stage. With respect to the pelvic angle, the *P. pyriformis* group is retarded, *P. clavatus* most of all.

In all groups except *P. pyriformis* development of the vault is accelerated in height and width over the pelvis. The top of the deltoids, the radials at the bases of the ambulacra, and the middle parts of the basal plates were contiguous in the nepionic stage (Pl. 13, fig. 1). The amount of addition of hard parts on both sides of the sutures is responsible for the different proportion of parts of the calyx, as well as for the difference in size of specimens. Those differences in proportion, if the proportions remain fairly constant, are responsible for specific differences. Less usual and less constant differences of proportion of parts, occurring within the species, are responsible for varieties, as the varieties of *P. godoni* and *P. conoideus*.

Differences in proportion of parts occur not only in ontogenetic stages but between individuals of the same species and the same varieties. It is therefore difficult to distinguish individual differences from varieties, varieties from species, and species from species, and indeed, to distinguish the genus *Pentremitidea* from *Pentremites*. Many varieties are probably only individual differences, such as *P. godoni major* and *P. conoideus perlongus*; they are allowed to stand, and other species are reduced to varieties, because it is convenient to designate a form by some appropriate name. On the other hand, specific names for ontogenetic stages—such as "*P. benedicti*" which is the nepionic stage of *P. conoideus* and "*P. koninckanus*" which is the neanic stage of *P. conoideus*—cannot stand.

*P. praematurus* is probably the young of *P. elegans*, and *P. turbinatus* the young of *P. pyriformis*. The ontogenetic stages of other species are not known as certainly as for *P. conoideus*.

The writers are not convinced that the larval stages of echinoderms described by Croneis and Geis (1940, p. 351) are those of *Pentremites princetonensis*, or even the young stages of any species of *Pentremites*. *Pentremites* of any ontogenetic stage do not have three equal basals but have two large basals, each made by the ankylosis of two equal plates, and one smaller plate—essentially five plates of equal size. *Pentremites* do not have arms or distal plates. The young individuals described by Croneis and Geis have no apparent bearing on the ontogeny of *Pentremites*. An ontogenetic series from larva to adult would be convincing, but the authors do not substantiate their statement that "the generic assignments are definite" (p. 346).

Plates 11 and 12 show ontogenetic series of *P. conoideus*, *P. godoni*, *P. girtyi*, *P. okawensis*, and *P. gutschicki*, and Hambach's series of *P. sulcatus*. The series shows several significant facts: (1) the pelvic angle increases during ontogeny for all groups of species except for representatives of *P. pyriformis*, *P. girtyi*, and *P. okawensis*; (2) the neanic stages tend to be globular and have sometimes been designated as the unrecognizable form "*P. globosus*".

The retention of the small pelvic angle of the nepionic stage in *P. girtyi*, *P. okawensis*, and *P. clavatus* is a good example of the biologic principle of retardation. It is not a return to ancestral condition, atavism, but the retention of an early ontogenetic character common to all pentremites.

## ANCESTRY OF *PENTREMITES*

Essie A. Smith noted (1906, p. 1235) that the early neanic, possibly nepionic, stages of *P. conoideus* are obconical, with a basal angle of 40° to 50°, oblique and short ambulacra, and "loose and oblique arrangement of poral pieces"; all these structures are very similar to those of *Codaster* from the Hamilton group of Ontario. She also suggests that "*Codaster* may be the ancestor of *Pentremites*, or *Codaster* and *Pentremites* may have descended from a common ancestor." In the young specimens of all species, in which young stages occur with adults, the pelvic angle of the young is less than that of the adult, except in retarded species of the *P. pyriformis* group. The vault is correspondingly shorter, and the ambulacral areas are shorter and more oblique to the axis of the calyx. Ulrich (1917, p. 244) noted the same tendency: "Younger specimens from the same locality have (like figs. 8 and 9) relatively more obconical bases; and in yet older examples the common tendency among *Pentremites* to reduce the height of the base is carried on to the extreme illustrated in figures 12 and 13 of *P. princetonensis*." (Pl. 4, figs. 5, 6)

The young stages of a species of *Pentremites* do not indicate what the immediate ancestral species was. All nepionic stages are obconical, and the neanic stage of most species is globular. The immediate ancestral species can be best determined by comparison of adult structures, especially the form of the calyx and the depth of the ambulacra, and by geologic horizon, using immature stages as a check. The conical nepionic stages of all species of *Pentremites* indicate that the ancestral genus was a conical form. The law of recapitulation applies to *Pentremites*, as it does to many groups of organisms. *Pentremitidea* is probably the immediate ancestor of *Pentremites*, and *Codaster* may be the ancestor of *Pentremitidea*. Some primitive crinoid, such as *Pisocrinus* or *Haplocrinites*, may have evolved into *Pentremites* by losing the arms and developing the ambulacral structures. The genus apparently did not evolve into any other form.

## PHYLOGENY OF *PENTREMITES*

The evolution of *Pentremites* within the group, or phylogeny, may be approximated, as with other invertebrates, by (1) comparison of similar structures or comparative anatomy, (2) application of the law of recapitulation, and (3) determining the geologic occurrence and applying the proposition that earlier forms are the ancestors of later forms. None of these methods is entirely reliable in determining the species to species phylogeny of the *Pentremites*.

Comparison of structures, check of geologic range, and an appreciation of ontogenetic stages have proved fairly satisfactory in making phylogenetic trees showing probable relationships.

On the basis of comparative structure of adult specimens, the pentremites are readily divisible into four groups: (1) the *P. conoideus* group, with very short pelves and convex narrow ambulacra, confined to the Lower and Middle Mississippian; (2) the *P. godoni* group, with short pelves, convex to flat ambulacra and short deltoids, characteristic of the lower Chester, but extending from the Ste. Genevieve to the upper Chester and possibly to the Pennsylvanian; (3) the *P. pyriformis* group, with long pelves, angles from 35° to 115°, nearly flat ambulacra and short deltoids, characteristic of the middle Chester; (4) the *P. sulcatus* group, with deep ambulacra and long deltoids, short pelves, and large size, characteristic of the Glen Dean, but extending from the Paint Creek formation to the top of the Chester Series (Fig. 1).

All species of *Pentremites* except the *P. pyriformis* group have smaller pelvic angles in the young stages. The *P. pyriformis* group retains the small-angled pelvis throughout life (which is not reversion to the obconical ancestral genus). It is necessary to have adult specimens to determine to which groups they belong by the angle of the pelvis and to distinguish species and varieties. Notwithstanding the decrease in size of the pelvic angle from the Lower and Middle Mississippian to the Upper Mississippian, and conversely, the paradox of its increase during the life history of individual pentremites, the pelvic angle is one of the two most reliable characters for separating all groups of *Pentremites* and for identifying species. The other most important character for identification is the convexity, flatness, or concavity of the ambulacra.

The *P. conoideus* group, the oldest group and structurally consistent, is ancestral to the *P. godoni* and *P. pyriformis* groups. The *P. godoni* group evolved from *P. conoideus* by flattening and widening of the ambulacra. The *P. pyriformis* group came from *P. conoideus* by retention of the small pelvic angle of the neanic stage and by flattening and widening of the ambulacra. The *P. sulcatus* group came from a form with short pelvis and slightly concave ambulacra; in the Golconda formation it arose from the *P. godoni* group. *P. platybasis* of the lower Chester fits all the requirements as ancestor of the *P. sulcatus* group. Figure 1 shows the geologic occurrence and phylogeny of the genus *Pentremites* and its four groups.

More detailed diagnoses of the characters of each group of species, with phylogenetic trees and keys to the groups, are given in Part 2. Under each group is a phylogenetic tree, which emphasizes the structural similarity of species and their stratigraphic occurrence.

## ECOLOGY OF *PENTREMITES*

Pentremites apparently lived only in clear water, where limestones or limy sediments were accumulating; none has been reported from sandstones, and only one species, *P. decussatus*, has been reported from clay shale. The limestones may have been silty (parts of the Golconda and Beech Creek limestones), coquina (parts of the Salem limestone), or oölitic (parts of the Ste. Genevieve and Beaver Bend limestones). The writers know of no pentremites from fine-grained or lithographic limestones, like the Levias member of the Ste. Genevieve limestone or much of the St. Louis limestone. Pentremites are also found in calcareous shales (the Paint Creek and Renault formations), but occurrence in shale is less common than for limestones.

The Harrodsburg limestone is very silty in its lower part, indicating an unfavorable environment of deposition, since pentremites are not present. The upper Harrodsburg limestone is much purer, and parts of it are like the Salem limestone in that there are beds of coquina in which fossils occur in abundance. Pentremites are abundant in the upper Harrodsburg limestone, continuing into the Salem limestone.

The coquinalike parts of the Salem and Harrodsburg limestones indicate that the depositional currents must have moved rapidly, for the fauna consists of comminuted large fossils plus the small fossils. The environment in which the Salem limestone was deposited was exceptionally favorable for life, for Foraminifera, corals, Bryozoa, gastropods, crinoids, brachiopods, and shark teeth and bones, as well as pentremites, are present in abundance. The presence of plants is not indicated except by the organisms which must have used great quantities of plants for food. The great number of organisms substantiates the idea of a shallow tropical sea near a low-lying land mass.

Oölitic beds in limestones commonly contain pentremites. Many of the species Ulrich recorded in 1917 had representatives in the Gasper and Fredonia oölites of Kentucky, Tennessee, and Alabama, and some specimens from the Beaver Bend limestones at Silverville are from an oölitic zone. This indicates considerable agitation of the water, since oölites do not form in a calm environment.

The Beech Creek limestone is in places a coarse crystalline coquina with several red sandy zones in which pentremites are found. These are thinner than the silty zone of the Harrodsburg, and a possible explanation for finding pentremites in one and not the other is that the pentremites could withstand the deposition of sand and silt for only a short time.

Pentremites occur in profusion in the Paint Creek formation of Illinois, which is a limy shale. The pentremites from this formation are exceptionally well preserved, but the pinnules and covering structures of the ambulacra are missing. In several hundred specimens from the Paint Creek formation, not even as coarse a structure as the stem is preserved, indicating deposition in agitated water. Allen and Lester (1953, p. 195) found the very large *P. maccalliei* (considered by them to be a new species) in an argillaceous limestone, and they think the muddy environment induced the large size; this is a doubtful conclusion, for the species is large in whatever kinds of rock it is found, and other species in shales are small.

Pentremites are very rare in black shales; several specimens in the U. S. National Museum preserved in black shale have stems but no ambulacral structures finer than grooves and ridges. The pentremites are not found in any pure sandstones. The Chester sandstones of Indiana are barren of fossils, and no occurrences of pentremites have been reported from pure sands.

In general, the habitat of the pentremites seems to have been shallow, warm, clear, agitated marine water, with a great number of other organisms.

## FOSSILIZATION OF *PENTREMITES*

The calyx wall or skeleton of *Pentremites* was composed of a meshwork of calcite rods or spicules. The wall was light and porous, much as it is for modern echinoderms. The skeletons of all pentremites have been altered in some way and changed to typical fossils. During fossilization water-bearing calcium carbonate moved through the rock, infiltrated the hollow spaces in the walls, and precipitated calcite, adding both weight and strength to the calyx. The calcite of the fossilized calyx and stem is in the form of large crystals, much as if the calyx were carved from a single crystal of calcite, indicating that the calcite meshwork was crystallized in the living animal, with the axes of calcite in their normal position.

The best-preserved calices occur in shale and show the growth lines and finer structures on the ambulacra; some still have the pinnules and structures over the mouth. Well-preserved specimens are found in the Renault shales, Paint Creek shales, shales in the Golconda and Glen Dean formations, and shales in the Menard, Clore, and Kinkaid formations.

Inside and outside the calyx were the soft parts which decayed. The only internal hard parts capable of preservation were the calcareous hydrospires. They were connected to the outside of the calyx by the spiracles at the summit and by the pores along the ambulacra. The solutions entered the calyx through the mouth, spiracles, and pores, as well as through the porous walls, and calcite precipitated out, filling the calyx and infiltrating the walls. Specimens of *P. angularis* Lyon from Cloverport, Kentucky, show perfect preservation of the hydrospires and filling of the calyx with clear calcite. Specimens of *P. rusticus* have hydrospires perfectly preserved and the calyx filled with finely comminuted, argillaceous, and calcareous mud, brought in through the mouth. The walls are infiltrated with calcite. Specimens from the Paint Creek formation of Illinois are filled with mud, and the walls are infiltrated with calcium carbonate.

Many specimens from the Salem and Harrodsburg limestones have an excess of calcite coating the specimens, obscuring the fine structures. Where the currents moved rapidly specimens are somewhat worn on the surface and even broken, as are specimens from "Pentremite Hollow", 2 miles south of Bloomington, Indiana. Fragments of pentremites occur in abundance in the Salem limestone at Paynters Hill, 4 miles southeast of Salem, Indiana, indicating strongly agitated water during deposition of the fossils. The type of *P. decussatus* is only a fragment from the New Providence shale, probably from a specimen broken in weathering.

Crystallization accompanies infiltration, especially in the echinoderms; the entire fossil then acts as a single calcite crystal or a few crystals. Crystallization is a hindrance to obtaining good specimens. Where part of a pentremite is exposed on a slab of limestone, attempts to extract it may cause cleavage across the structures. Some infiltrated specimens are hollow, with calcite crystals attached to the walls, simulating geodes.

A common type of fossilization in specimens preserved in limestones is silicification. During weathering the limestone is exposed to the action of ground water carrying silica in solution. The silica then tends to replace the calcite, and destroys much fine detail on many specimens. Many hundreds of specimens from the Golconda formation near Grantsburg, Indiana, are completely silicified and weathered hollow. Most of the Glen Dean specimens from Cloverport, Kentucky, Rome, Indiana, and Crane, Indiana, are silicified but retain the calcite filling, with the limestone still attached outside on many specimens. In some specimens, such as those of *P. hambachi* from the Golconda formation, the walls of the calyx are calcite, and the material inside is chert which has destroyed the hydrospires. Silicification may affect the entire specimen including the cavity filling, as in *P. spicatus* from Cloverport, Kentucky. Silification may also occur as "Beekite" rings, which distort fine structures.

The last stage is geodization, as in *P. godoni* from Warren County, Kentucky (Pl. 13, figs. 9–12). Silica deposits in the fractures and expands in crystallizing to quartz, producing more fractures and cracks. By repeated deposition and crystallization, the fossil enlarges, becomes hollow, and becomes a typical geode, with a shell of chalcedony and an inner layer of quartz or chalcedony. Eventually the geodes become so large that all traces of the organisms in which they began to form are lost.

Much of the limestones in which the pentremites are found weathers to a residual soil; it may be profitable to sift some of the soil for specimens. Pentremites such as those from the Golconda formation, Grantsburg, Indiana, and the Salem limestone of Paynters Hill, 4 miles southeast of Salem, Indiana, are generally colored red by the iron oxide in the soil, and many are pitted and may be hollow because of solution. These specimens are fragile, and most of the fine structures are lost by weathering.

Some specimens are more crushed in the ambulacra, reducing flat ambulacra to a concave shape. Since the shape of the ambulacra is important in identification, such distortion must be taken into consideration. The deltoids and ambulacral flanges are broken or weathered off in many specimens; in these specimens it would be difficult to separate such forms as *P. halli* and *P. angularis*, in which those structures are important for identification.

Many pentremites are crushed by the weight of overlying sediments, before infiltration by lime, and are later infiltrated, as are most of the specimens of *P. tulipaformis* and *P. gutschicki* from the Kinkaid formation, 8 miles northeast of Marion, Kentucky. Large specimens of *P. altus* from the Kikaid formation, 2 miles west of Robbs, Illinois, have been crushed before infiltration, emphasizing the lightness of the calyx, rather than the solid character of bone. Most specimens of the large *P. maccalliei* have been distorted by compaction of the sediment in which they were buried.

Some specimens have been involved in actual diastrophic movements, faulting and folding, so that they are distorted, as are the specimens of *P. rusticus* from Arkansas.

Internal molds, such as those figured by Etheridge and Carpenter (1886, Pl. 3, figs. 10–12) are of small value in identification, although they give some information concerning internal structure. Casts or molds, such as those illustrated by Butts (1941, Pl. 132, figs. 26, 27), leave much to be desired for description and identification.

# PART 2. SYSTEMATIC DESCRIPTIONS OF GENUS AND SPECIES

## CLASSIFICATION OF THE GENUS *PENTREMITES*

Phylum     ECHINODERMATA Laske, 1878
Subphylum PELMATOZOA Leuckart, 1848
Class      BLASTOIDEA Say, 1825
Order      EUBLASTOIDEA Bather, 1899
Family     PENTREMITIDAE d'Orbigny, 1852
Genus     **Pentremites Say, 1820**

Genotype (designated by Etheridge and Carpenter, 1886, p. 157), *Pentremites godoni* (Defrance). (*Pentremites florealis* Say, 1825 = *Encrinites florealis* von Schlotheim, 1820 = *Encrina Godonii* Defrance, 1819) (Chester formation, Kentucky)

"An Asterial Fossil", PARKINSON, 1808, Organ. Rem., vol. 2, p. 235, pl. 13, figs. 36–37
"Echinus of the family galerite", MITCHILL, 1818, *in* Cuvier's Theory of the Earth, New York, p. 363, pl. 3, fig. 6
*Encrina Godonii* DEFRANCE, 1819, Dict. Sci. Nat., vol. 14, p. 467
*Encrinites florealis* VON SCHLOTHEIM, 1820, Petrefactenkunde, p. 339 (also given as vol. 2, p. 38)
*Pentremite* SAY, 1820a, Amer. Jour. Sci., vol. 2, no. 2, p. 36.—SAY, 1825, Jour. Acad. Nat. Sci., Philadelphia, 1st ser., vol. 4, no. 9, p. 291. Reprint, 1826, Zool. Jour., vol. 2, no. 7, London, p. 311. Reprint, Bull. Amer. Pal., 1895, p. 284, 350
*Pentremites* SOWERBY, 1826, Zool. Jour., vol. 2, no. 7, London, p. 316.—ETHERIDGE AND CARPENTER, 1886, Cat. Blastoidea Geol. Dept. British Mus. (Nat. Hist.), London, p. 150.—WELLER, 1898, U. S. Geol. Surv., Bull. 153, p. 411 (Index).—GRABAU AND SHIMER, 1910, North American Index Fossils, vol. 2, A. G. Seiler and Co., New York, p. 481, text figs.—WACHSMUTH, 1900, *in* Zittel-Eastman Textbook of Paleontology, Macmillan and Co., New York, p. 188.—SPRINGER, 1913, *in* Zittel-Eastman Textbook of Paleontology, Macmillan and Co., London, p. 161.— TWENHOFEL AND SHROCK, 1935, Invertebrate Paleontology, McGraw-Hill Book Co., Inc., New York, p. 167.—SHIMER AND SHROCK, 1944, Index Fossils of North America, Mass. Inst. Tech. Press, John Wiley and Sons, Inc., New York, p. 133.—MOORE, LALICKER, AND FISCHER, 1952, Invertebrate fossils, McGraw-Hill Book Co., Inc., New York, p. 594–603, figs. 17–8, 9, 10.—SHROCK AND TWENHOFEL, 1953, Principles of Invertebrate Paleontology, McGraw-Hill Book Co., Inc., New York, p. 660, figs. 14–9A–G
*Pentatremites* GOLDFUSS, 1826, Petrefacta Germaniae, Theil 1, p. 160
*Pentatrematites* ROEMER, 1851, Archiv. f. Naturgesch., Jahrg., 17, Bd. 1, p. 324, 353

Calyx ovoid, pyriform, or globose, less commonly clavate, conoid, or bulbous; greatest width basal to supramedian, commonly median; basal periphery pentagonal to stellate; pelvis pentagonal, never three-sided; basal plates three, small, two pentagonal, one quadrangular, forming a shallow inverted cone; radial plates five, comprising more than half of calyx, each notched by a deep sinus in which ambulacrum occurs, the two limbs of each truncated terminally and across the radial sutures, visible from the side; deltoids five, small, four-sided, located supraradially, always less than half the length of calyx; ambulacra broad to narrow, elevated or flat to greatly depressed, composed of lancet, sublancet, and side plates, divided equally by narrow, depressed, median food groove, and transversely by numerous small lateral food grooves; round or oval hydrospire pores occur on side or outer side plates, generally visible only on weathered specimens or in sections; tenuous brachioles attached to side plates in exceptionally well-preserved specimens; 10 folded lamellar hydrospires below ambulacra; one fascicle of from four to six tubes beneath and parallel to each side plate and connecting with the hydrospire pores; spiracle openings five, four equal in size and divided by a plate into two openings, which may appear as double pores depending on degree of weathering, the fifth larger with a third opening containing the anus; mouth stellate, located at center of summit within the ring of spiracles; mouth and spiracles covered by small plates or rods in well-preserved specimens, place of stem attachment basal, marked by flattened area with central pore.

Lower to Upper Mississippian, possibly into the Lower Pennsylvanian. Confined to the United States and Canada, except for one doubtful species supposedly from the Devonian of Germany.

## GUIDE FOR DESCRIPTION OF SPECIES OF *PENTREMITES*

The following Guide for Description of Species of Pentremites was used in comparing all pentremite material, including actual specimens, photographs, drawings, and written descriptions. The characters listed are available from at least one of the sources named. After one becomes familiar with the pentremites, it is possible to identify species without seeing the specimen by enumerating the measurements listed on the guide. The more supplementary material one has, the more reliable is his identification. The Guide is designed for mature specimens, since young stages are generally considerably different from the adult—*e.g.*, "*P. koninckanus*" and *P. conoideus*.

Species
Plates and figures
Synonymy, with horizon and locality
Calyx: shape, length, width, position of greatest width, L/W
Vault: shape, sides and summit, length
Pelvis: length, V/P, shape of sides, pelvic angle
Ambulacra: convexity or concavity, rims, transverse grooves in 3 mm
Interambulacra: convexity, flatness, or concavity
Deltoids: length, below, even with or above spiracles, flare
Geologic range of species, localities, and horizons
Special points about the species

## GROUPS OF SPECIES OF *PENTREMITES*

Division of the 63 recognizable species of *Pentremites* into four groups of species was necessary to determine the geologic range and evolution of the genus, and it facilitates identification of species and determination of their genetic relationships.

It is not desirable to make separate genera, or even subgenera, for the four groups; nothing would be gained by the multiplication of genera. It would be possible to make at least 15 groups of species, as is actually done in both phylogenetic trees and in the keys.

Keys are provided for the groups of species and for the species under each group. The most reliable and usable characters for identification are: length-width ratio, vault-pelvis ratio, pelvic angle, convexity or concavity of the ambulacra, and the degree of concavity of the interambulacra. To those five characters are added, for the *P. sulcatus* group, the protrusion of the deltoids above the mouth and spiracles.

If two or more species or varieties cannot be separated by a single constant character, they are considered synonyms of previously named species; there are 24 named species and varieties which are synonyms of prior species. A score more species might have been sacrificed. On the other hand, many specimens, perhaps real species, may be found which cannot be identified by means of the keys, and students will name them new species.

## 1. *PENTREMITES CONOIDEUS* GROUP

Calices of medium size, ovoid to pyramidal or conoidal; vault 4–20 times height of pelvis; basal angle 130°–180°; ambulacra narrow, 2–3 mm, convex; ambulacral rims not flanged; interambulacra concave; deltoids end below spiracles. Lower and Middle Mississippian, abundant in the upper Harrodsburg and Salem limestones (Table 2; Fig. 2).

## 2. *PENTREMITES GODONI* GROUP

Calices of medium size, subglobular to conoidal; vault 3–20 times pelvis; basal angle 110°–170°; ambulacra wide, 3–5 mm, slightly convex to flat (slightly concave at the top in one species); ambulacral rims low and narrow; interambulacra slightly concave; deltoids end below spiracles. Ste. Genevieve to Lower Pennsylvanian?, common in the lower and middle Chester (Table 3; Fig. 3).

## 3. *PENTREMITES PYRIFORMIS* GROUP

Calices small to large, ovate, pyriform to clavate; vault 0.5 to 3 times pelvis; basal angle 35°–110°; ambulacra wide, 3–8 mm, slightly convex to flat; ambulacral rims low and narrow; most interambulacra nearly flat; deltoids below spiracles. Ste. Genevieve and Chester, abundant in the middle Chester, rare in the upper Chester. (Table 4; Fig. 4).

## 4. *PENTREMITES SULCATUS* GROUP

Calices, medium to very large, ovoid; vault 0.6 to 7 times pelvis; basal angle 80°–165°; ambulacra wide, 5–10 mm, and deep, 2–5 mm, ambulacral flanges moderate to strong; interambulacra moderately to strongly concave; deltoids even with or above spiracles. Lower Chester rare, middle Chester common, upper Chester few species (Table 5; Fig. 5).

### KEY TO GROUPS OF *PENTREMITES*

Page

1a. Ambulacra convex to flat; deltoids end below spiracles
  2a. Basal angle 110°–180°
    3a. Ambulacra strongly convex, narrow; interambulacra strongly concave—*P. conoideus*
      group.................................................................... 39
    3b. Ambulacra slightly convex to flat, broad; interambulacra nearly flat—*P. godoni* group... 45
  2b. Basal angle 35°–110°—*P. pyriformis* group...................................... 51
1b. Ambulacra concave; deltoids reach or extend above spiracles—*P. sulcatus* group......... 61

### KEY TO SPECIES OF *PENTREMITES CONOIDEUS* GROUP

1a. Greatest diameter above base of ambulacra; calyx ovoid
  2a. Ambulacra very convex, narrow, 2 mm
    3a. Side plates cover lancet plate—*Pentremitidea leda* (Hall)............................ 40
    3b. Side plate does not cover lancet plate
      4a. Deltoids short, 2–3 mm—*Pentremites decussatus* Shumard....................... 41
      4b. Deltoids long, 5–7 mm—*P. conoideus amplus* Rowley........................... 44
  2b. Ambulacra gently convex, moderately wide, 3–5 mm
    3c. Basal angle over 150°
      4c. Length to width, 1.1; summit convex—*P. burlingtonensis* M. & W.............. 42
      4d. Length to width, 1.4; summit flat—*P. elongatus* Shumard.................... 42
    3d. Basal angle 130°
      4e. Sides of pelvis straight—*P. ovoides* Ulrich................................. 44
      4f. Sides of pelvis concave—*P. ovalis* Goldfuss................................ 44
1b. Greatest diameter at base of ambulacra
  2c. Calyx longer than wide
    3e. Calyx conoidal; length over width, 1.3—*P. conoideus* Hall...................... 42
    3f. Calyx cylindroidal; length over width, 1.5—*P. conoideus perlongus* Rowley.......... 43

Page

2d. Calyx length and width subequal
    3g. Sides of vault moderately convex, greatest width at base—*P. conoideus obtusus* Ham-
        bach...................................................................................... 43
    3h. Side of vault strongly convex, greatest width median—*P. conoideus amplus* Rowley... 44
        syn. *P. cavus* Ulrich............................................................... 44

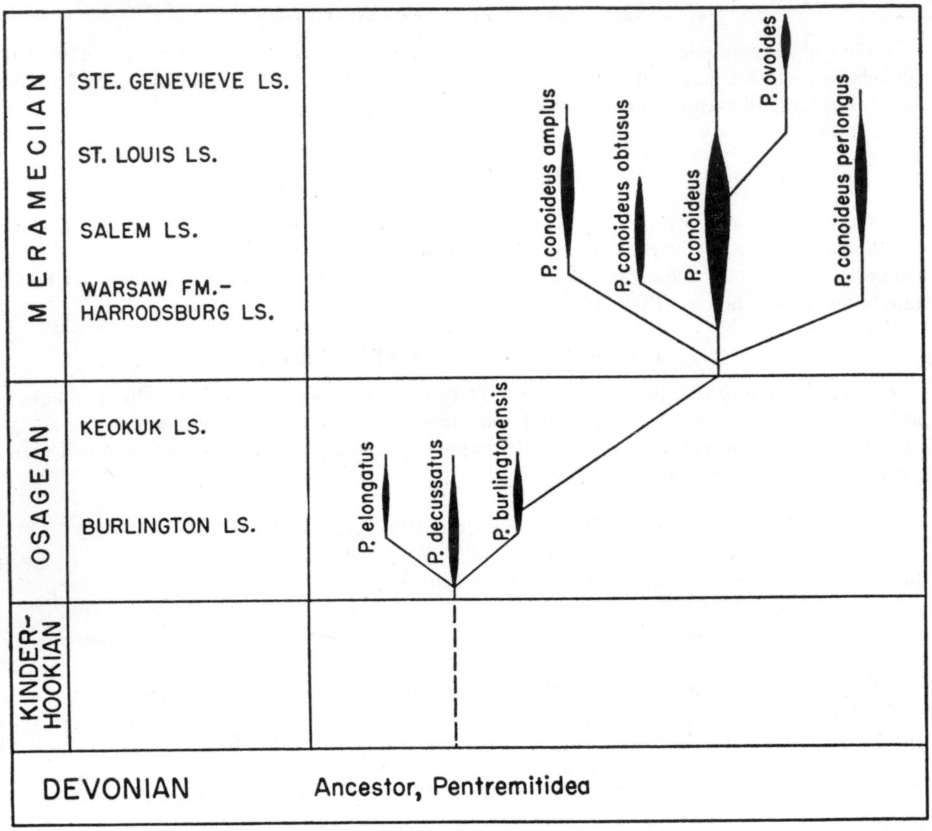

FIGURE 2.—*Geologic occurrence and phylogeny of the* Pentremites conoideus *group*

## DESCRIPTIONS OF SPECIES OF *PENTREMITES CONOIDEUS* GROUP

### Pentremitidea leda (Hall)

(Pl. 2, fig. 1)

*Pentremites leda* HALL, 1862, State Cabinet Nat. Hist. N. Y., 15th Rept., p. 149, Pl. 1, fig. 11 (Middle Devonian, Hamilton group, western New York). Type species of *Devonoblastus* Reimann, 1935

*Pentremitidea? leda*, ETHERIDGE AND CARPENTER, 1886, Cat. Blastoidea Geol. Dept. British Mus. (Nat. Hist.), Pl. 5, figs. 12–14 (Middle Devonian, Ontario)

Calyx ovoid, length 22 mm, diameter 11 mm; greatest width submedian, above ambulacral tips; L/W, 2; pelvis short, about 3 mm; V/P, 7; basal angle, 120°–130°; ambulacra long, narrow, convex, whether lancet plates or only side plates visible in doubt; interambulacra concave; deltoids short, 2 mm; summit truncate; "summit openings small"; ambulacral grooves 9 in 3 mm.

Middle Devonian, Hamilton of western New York and Ontario.

This genus and perhaps the species seem to be the ancestor of *Pentremites*. It is much like *P. conoideus* of the Salem limestone, differing in the shorter deltoids and less-exposed lancet plates. It is important to note the short base, rather than the elongate base which is expectable on the theory that the immediate ancestral form should have a conical pelvis, as do the young stages of *Pentremites*. The type species of *Pentremitidea* does have a conical pelvis.

TABLE 2.—*Main characters of the* Pentremites conoideus *group*

Calices ovoid to conoidal, of medium size; vault 4–20 times pelvis; basal angle 130°–180°; ambulacra narrow, 2–3 mm, convex, with low rims; interambulacra concave; deltoids below summit. Lower and Middle Mississippian. (Synonyms in italics)

| Species | Length / Width | Vault / Pelvis | Pelvic angle | Ambulacra | Interambulacra | Original horizon and locality |
|---|---|---|---|---|---|---|
| P. decussatus | 1.1 | 6 | 180° | convex | concave | New Providence; Ky. |
| P. burlingtonensis | 1.1 | 5 | 125°–155° | convex | slightly concave | Upper Burlington; Iowa |
| P. elongatus | 1.5–1.8 | 6–20 | 160° | convex | slightly concave | Upper Burlington; Mo. |
| P. conoideus | 1.2–1.4 | 4–8 (5–6) | 120°–150° av. 135° | narrow convex | deeply concave | Salem; Ind. |
| *P. benedicti* (nepionic P. conoideus) | 1.3 | 0.8 | 50°–65° | flat | flat | Warsaw; Ill. |
| *P. koninckanus* (neanic P. conoideus) | 1.1 | 1 | 80° | slightly convex | concave | Salem; Ind. |
| P. conoideus perlongus | 1.5 | 20 | 140°–180° | narrow convex | deeply concave | Salem; Ind. |
| P. conoideus obtusus | 0.9 | 16 | 180° | narrow convex | slightly concave | Warsaw; Ill. |
| P. conoideus amplus | 1 | 14 | 180° | narrow convex | deeply concave | Salem; Ind. |
| *P. cavus* | 1 | 18 | 180° | narrow convex | deeply concave | St. Louis; Ky. |
| P. ovoides | 1.3–1.4 | 6–10 | 130° | convex | slightly concave | Ste. Genevieve; Ky. |
| P. ovalis | 1.5 | 5 | 130° | concave? | concave below | Uncertain; Germany |

## Pentremites decussatus Shumard

(Pl. 2, figs. 2, 3)

*Pentremites decussatus* SHUMARD, 1858, Trans. St. Louis Acad. Sci., vol. 1, pt. 2, p. 242, Pl. 9, fig. 6 (Lower Mississippian, New Providence shale, Button Mould Knobs, 7 miles south of Louisville, Ky.).—WELLER, 1909, Geol. Soc. Am., Bull., vol. 20, p. 288, Pl. 11, figs. 28, 29 (Fern Glen formation, Fern Glen, Mo.)

Calyx globular, 20 mm long and 18 mm wide, with greatest width submedian; L/W, 1.1; vault 17 mm long, with moderately curved sides and nearly flat summit; pelvis 3 mm long, with convex sides and with basal plates in a depression; V/P, 6; pelvic angle near 180°; ambulacra 17 mm long, narrow, 2 to 3 mm wide, strongly convex, without raised rims; lancet plates nearly covered by side plates; interambulacra concave; about 5 transverse ridges in 3 mm; deltoids short, 2–3 mm, just reaching the summit.

Lower Mississippian, New Providence shale, Ky.; Fern Glen formation, Mo.

This species has all the characters of typical members of the *P. conoideus* group. It is very similar to *P. conoideus amplus* and differs mainly in that it is relatively longer. It is the oldest known species of the genus (excluding *P. ovalis* Goldfuss).

### Pentremites burlingtonensis Meek and Worthen

#### (Pl. 2, fig. 4)

*Pentremites Burlingtonensis* MEEK AND WORTHEN, 1870, Proc. Acad. Nat. Sci. Philadelphia, p. 33.—
    1873, Geol. Surv. Ill., vol. 5, p. 461, Pl. 8, fig. 7 (Upper Burlington limestone, Burlington, Iowa)
*Pentremites burlingtonensis* WELLER, 1898, U. S. Geol. Surv., Bull. 153, p. 412

Calyx short melon-shaped, length 24 mm, width 21 mm, median; L/W, 1.1; vault and summit evenly convex; vault long, 20 mm; pelvis short, 4 mm, with nearly straight sides; V/P, 5; basal angle 125°–155°; ambulacra convex, with low rims; about 8 transverse ridges in 3 mm; interambulacra slightly concave; deltoids 7 mm long, not reaching the summit; the four spiracles are double.

Upper Burlington limestone, Burlington, Iowa.

The species is rare; Meek and Worthen had seen only three specimens, and it has not been again described. The double spiracles at the surface are unusual.

### Pentremites elongatus Shumard

#### (Pl. 2, figs. 5–7)

*Pentremites elongatus* SHUMARD, 1855, Geol. Surv. Mo., 2nd Ann. Rept., p. 187, Pl. B, fig. 4 (Upper
    Burlington, Clarksville, Columbia, Rocheport, Mo.).—ETHERIDGE AND CARPENTER, 1886, Cat.
    Blastoidea Geol. Dept. British Mus. (Nat. Hist.), p. 161, Pl. 1, figs. 4, 5; Pl. 2, figs. 14, 15;
    Pl. 18, fig. 4.—KEYES, 1894, Geol. Surv. Mo., vol. 4, pt. 1, p. 133, Pl. 18, fig. 4 (Upper Burling-
    ton limestone, Burlington, Iowa)

Calyx subcylindrical, 31 mm long, 19 mm wide; greatest width median; L/W, 1.4; sides of vault slightly convex; summit flat; length of vault 29 mm; V/P, 15; pelvis 2 mm long with nearly straight sides; pelvic angle, 160°; ambulacra convex, 4–5 mm wide, without raised rims; interambulacra slightly concave; about 5 transverse ridges in 3 mm; deltoids 10 mm long, nearly reaching summit.

Upper Burlington limestone, Clarksville, Columbia, Rocheport, and other places in Missouri.

The figure and specimens of Etheridge and Carpenter and of Keyes are elliptical and more like *P. burlingtonensis* and *P. ovoides* than Shumard's figure of *P. elongatus*. Shumard's figure is similar to *P. rusticus* but is much longer (and *P. rusticus* is in the supposed Lower Pennsylvanian).

### Pentremites conoideus Hall

#### (Pl. 2, figs. 8–15; Pl. 11, figs. 1–19; Pl. 13, fig. 8)

*Pentremites conoideus* HALL, 1856, Trans. Albany Inst., vol. 4, p. 5, (Salem limestone, Spergen Hill
    and Bloomington, Ind.).—1858, Rept. Geol. Surv. Iowa, vol. 1, pt. 2, p. 655, Pl. 22, figs. 9, 10
    (Fig. 8 is var. *perlongus* Rowley) (Same localities).—WHITE, 1881, 2nd Ann. Rept. Dept.
    Statistics and Geol. Ind., p. 512, Pl. 7, fig. 12 (Spergen Hill, Bloomington, and other places in
    Indiana).—HALL, 1883, 12th Ann. Rept. Dept. Geol. Nat. Hist., p. 323, Pl. 32, fig. 32 (Spergen
    Hill, Lanesville, and Bloomington).—ROWLEY, *in* GREENE, 1902, Contrib. to Ind. Paleo., vol.
    1, pt. 10, p. 87, Pl. 29, figs. 29, 37–41.—ULRICH, 1905, U. S. Geol. Surv., Prof. Pap. 36, p. 57,
    Pl. 6, figs. 1–6 (Lower St. Louis limestone, Princeton, Ky.).—SMITH, 1906, Ind. Dept. Geol.
    Nat. Res., 30th Ann. Rept., 1905, p. 1219–1242, Figs. 1–3, Pls. 43–48 (Upper Harrodsburg and
    Salem limestone, Harrodsburg, Bloomington, Spergen Hill and Paynters Hill in Washington
    Co., Ind., Ellettsville, Stinesville, and Lanesville. Discussion of structure, ontogeny, and sup-
    posed dwarfing).—PECK, 1930, Pan-Am. Geol., vol. 54, p. 106, Pl. 14, figs. 6–8 (Upper Mis-
    sissippian, Brazer limestone, Mendon, Utah)
*Pentremites koninckanus* HALL, 1856, Trans. Albany Inst., vol. 4, p. 4.—1858, Rept. Geol. Surv.
    Iowa, vol. 1, pt. 2, p. 656, Pl. 22, figs. 1a, b, c (Salem limestone, Spergen Hill and Blooming-
    ton).—WHITFIELD, 1882, Bull. Amer. Mus. Nat. Hist., vol. 1, p. 43, Pl. 3, fig. 33 (Salem lime-
    stone, Spergen Hill, Bloomington, Ind., Alton, Ill.).—BEEDE, 1906, Ind. Dept. Geol. Nat.
    Res., 30th Ann. Rept. for 1905, Pl. 26, fig. 33.—SMITH, E. A., 1906, Ibid., Pls. 45–47 (Neanic
    stage of *P. conoideus*)
*Pentremites benedicti* ROWLEY, 1900, Am. Geol., vol. 25, p. 69, Pl. 2, figs. 29–32. (Warsaw limestone,

Grand Tower, Ill.; Wittenberg, Mo.; Harrodsburg or Salem limestone, Ind.) (Nepionic stage of *P. conoideus*, younger than the neanic stage, *P. koninckanus*)

*Pentremites downeyensis* ULRICH, 1917, Miss. Form. West. Ky., Ky. Geol. Surv., p. 170 (Late Ste. Genevieve limestone)

Calyx conoidal with narrow summit, up to 22 mm long, av. adult 17 mm, width up to 17 mm, av. adult 13 mm; greatest width at points of ambulacra; L/W, 1.2–1.4, av. 1.3; vault with moderately curved sides, av. 13 mm long; pelvis 2–4 mm long, with nearly straight sides; V/P, 4–8, av. 5–6; pelvic angle, 120°–150°, av. 135°; ambulacra narrow, about 2 mm, strongly convex, with scarcely any rims; 7–8 transverse ridges in 3 mm, interambulacra deeply concave; deltoids 6–10 mm long, ending about 1 mm below summit.

Keokuk to Ste. Genevieve, abundant in the upper Harrodsburg and lower Salem limestones; Keokuk limestone, Boonville, Mo. (Keyes, 1894); Warsaw formation, Grand Tower, Ill. (Rowley, 1905), Blount Springs, Ala. (Butts, 1926), Keokuk, Iowa (Van Tuyl, 1925); upper Harrodsburg, Greenville, Paynters Hill, Washington Co., 1½ miles south of Harrodsburg, 1 mile north of Bloomington, Ind.; Salem limestone, Lanesville, Spergen Hill, Old Cleveland Quarry, Harrodsburg, Bloomington, Ellettsville, Stinesville, Ind.; Lower St. Louis limestone, Princeton, Ky. (Ulrich, 1905); Ste. Genevieve limestone, Pella, Iowa (Van Tuyl, 1925); Brazer limestone, Mendon, Utah (Peck, 1930); Carboniferous, Old Baldy, Mont. (Clark, 1917).

*P. conoideus* is one of the most easily recognized species of the genus. Most of the specimens vary little from the typical form, but perhaps one in 100 specimens is as wide as long (*P. obtusus* Hambach, *P. amplus* Rowley, *P. cavus* Ulrich), and perhaps one in 1000 is twice as long as wide (*P. conoideus perlongus* Rowley).

Whitfield, as well as Beede and Smith, considered *P. koninckanus* merely young specimens of *P. conoideus*. The constant association of the two forms, and the fact that perfectly graded series from small to large and from long pelvis and small angle to short base and high angle have been demonstrated by Smith, are convincing evidence that *P. koninckanus* is the young of *P. conoideus*. Smith (1906, p. 1237) considered that the small pelvic angle of the young of *P. conoideus* indicated that the ancestor of *P. conoideus* was a *Codaster*. Except for the small angle of the young of *P. conoideus*, there is little generic similarity of the two genera; *Codaster* lacks the round spiracles and the convex ambulacra of *P. conoideus*. Smith (1906, p. 1237) also considered that the specimens from the Old Cleveland Quarry, 1 mile north of Harrodsburg, had been dwarfed slightly; they have a breadth of 5.67 mm, compared with a breadth of 6.40 mm for specimens from Pentremite Hollow, 2 miles south of Bloomington. Dwarfing of the Salem fauna as a whole may be seriously questioned.

## Pentremites conoideus perlongus Rowley

### (Pl. 2, figs. 16–18)

*Pentremites conoideus perlongus* ROWLEY, *in* GREENE, 1902, Contrib. to Ind. Paleo,. pt. 10, p. 87, Pl. 29, fig. 28 (Upper Harrodsburg or Salem limestone, Lanesville, Ind.).—BEEDE, 1906, Ind. Dept. Geol. Nat. Res., 30th Ann. Rept. for 1905, p. 1263, Pl. 7, fig. 7 (Quotes from Rowley)

Calyx elongate, 19–21 mm long, 12–14 mm wide, greatest width near the base; L/W, 1.5; vault moderately convex, 18–20 mm long; pelvis very short, scarcely more than 1 mm, with nearly straight sides; V/P, 20; pelvic angle, 140°–180°; ambulacra narrow, very convex, with scarcely any rims; interambulacra deeply concave; 7–8 transverse ridges in 3 mm; deltoids 5–7 mm long, not reaching the summit.

Upper Harrodsburg and lower Salem limestone, Lanesville and Harrodsburg, Ind.

This variety is very rare; there are only three in more than 1000 specimens of *P. conoideus*.

## Pentremites conoideus obtusus Hambach

### (Pl. 2, figs. 19, 20)

*Pentremites obtusus* HAMBACH, 1903, Trans. St. Louis Acad. Sci., vol. 13, p. 53, Fig. 13 (Warsaw limestone, Boonville, Mo.)

Calyx short conoidal, typically wider than high, up to 19 mm wide and 16 mm high; greatest width at base; L/W, 0.9; vault with moderately curved sides and flat summit; pelvis nearly flat with sigmoid sides; V/P, 16; pelvic angle 180°; ambulacra long, narrow, convex, with low rims; interambulacra concave, "not as much depressed as we find it in *Pentremites conoideus*" (Hambach, 1903, p. 53); deltoids about 7 mm long, nearly reaching the summit.

Reported only from the "Warsaw" limestone of Boonville, Mo.; probably the upper Warsaw, or even the Salem limestone. Occurs fairly commonly in the upper Harrodsburg limestone, at Pentremite Hollow, 2 miles south of Bloomington, Ind., and 1½ miles south of Harrodsburg, Ind.

This variety differs only from the variety *P. amplus* in the less convex outline of the vault. It is obviously a variety of *P. conoideus*. The writers have about a dozen specimens from 1½ miles south of Harrodsburg, which are intermediate between *P. conoideus amplus* Rowley and *P. conoideus obtusus* Hambach. Neither variety is of any stratigraphic significance, but, with *P. conoideus perlongus* Rowley, they illustrate the main directions of variation within the species of *P. conoideus*.

### Pentremites conoideus amplus Rowley

#### (Pl. 2, figs. 21–26)

*Pentremites conoideus amplus* ROWLEY *in* GREENE, 1902, Contrib. to Ind. Paleo., pt. 10, p. 38, Pl. 29, figs. 31–34 (Upper Harrodsburg or Salem limestone, Lanesville, Ind.).—BEEDE, 1906, Ind. Dept. Geol. Nat. Res., 30th Ann. Rept. for 1905, p. 1263, Pl. 7, fig. 8
*Pentremites cavus* ULRICH, 1905, U. S. Geol. Surv., Prof. Pap. 36, p. 57, Pl. 6, figs. 7, 8 (Lower St. Louis limestone, Princeton, Ky.).—BUTTS, 1917, Miss. Form. West. Ky., Ky. Geol. Surv., p. 45, Pl. 10, figs. 18, 19 (After Ulrich)

Calyx as wide as long, up to 18 mm; greatest width basal or suprabasal; L/W, 1; vault and summit subcircular in outline; pelvis flat to 2 mm long, with straight sides; V/P, 14; pelvic angle, 150°–180°; ambulacra narrow and very convex with scarcely any rims; interambulacra very concave; deltoids 5–7 mm long, not reaching summit.

Upper Harrodsburg or lower Salem limestone, Lanesville and 1½ miles south of Harrodsburg, Ind.; Upper Harrodsburg limestone, 2 miles south of Bloomington, Ind.; lower St. Louis limestone, Princeton, Ky.

This variety occurs with typical *P. conoideus*, and there are all gradations between the two, although the variety is rare (scarcely one in 100 specimens of *P. conoideus*). *P. cavus* occurs a formation higher in the column, the St. Louis, but there appears to be no difference between it and *P. conoideus amplus*.

### Pentremites ovoides Ulrich

#### (Pl. 2, figs. 27, 28)

*Pentremites ovoides* ULRICH, *in* BUTTS, 1917, Miss. Form. West. Ky., Ky. Geol. Surv., p. 61, Pl. 15, figs. 6, 7 (Ste. Genevieve limestone, Livingston Co., Ky.)

Calyx stout, elliptical, 21–22 mm long, 15–17 mm wide, greatest width median; L/W, 1.4–1.3; vault evenly convex, 19–20 mm long, summit strongly convex; pelvis short, 2–3 mm with straight sides; V/P, 6–10; pelvic angle, 130°; ambulacra convex, with narrow, low rims; interambulacra slightly concave; deltoids 2 mm below summit, length not shown.

Known only from the Ste. Genevieve limestone of Livingston Co., Ky.

The form is much like that of the Burlington species. It was not described, and the figures are not good.

### Pentremites ovalis (Goldfuss)

#### (Pl. 2, fig. 29)

*Pentatremites ovalis* GOLDFUSS, 1826, Petrefacta Germaniae, Th. 1, p. 161; 2nd ed., 1862, p. 150, Pl. 50, figs. 1a, b, c (Transition limestone, Cromford, near Ratigen, Dusseldorf, Germany).—ETHERIDGE AND CARPENTER, 1886, Cat. Blastoidea British Mus. (Nat. Hist.), p. 156 (Doubted age and generic characters)

Calyx ovoid, length 9 mm, width 6 mm, greatest width submedian; L/W, 1.5; vault with straight sides and convex summit; pelvis short, with long stem; V/P, 5; sides concave; basal angle, 130°;

ambulacra wide, apparently concave, side plates and lancet plate not distinguished, ambulacral rims not shown; transverse grooves about 13 in 3 mm; interambulacra concave at base and convex at deltoids; deltoids not reaching summit, length not shown. Spiracles and mouth typical of *Pentremites*.

Transition limestone of Dusseldorf, Germany, supposedly Devonian.

The species has never again been described. Etheridge and Carpenter did not see Goldfuss' specimen and stated, "Its geological and its systematic position are therefore alike uncertain." It is the only species of *Pentremites* outside the United States and Canada and the only Devonian species. Its form is much like that of *P. ovoides*, *P. pulchellus*, and *P. tuscumbiae*, all of the Ste. Genevieve formation. Its Devonian age may be doubted, as well as its occurrence in Europe, inasmuch as another specimen has not been reported from Europe since 1826.

## KEY TO SPECIES OF *PENTREMITES GODONI* GROUP

Page

1a. Ambulacra connex to flat
  2a. Deltoids not flaring; calyx ovoid
    3a. Length equal to or greater than width
      4a. Length and width subequal; 1–1.1
        5a. Vault semicircular in side view—*P. godoni pinguis* Ulrich........................ 45
        5b. Vault paraboloid in side view—*P. godoni* (Defrance)........................... 48
            syn. *P. planus* Ulrich...................... 49
        5c. Vault a truncate pyramid—*P. malotti* n. sp............................... 48
      4b. Length much greater than width; 1.2–1.5
        5d. Basal angle 130°–160°
            6a. Vault tends to be parallel-sided—*P. godoni major* Etheridge and Carpenter..... 49
            6b. Vault tends to be conical—*P. godoni angustus* Hambach.................... 50
        5e. Basal angle 115°–130°—*P. tuscumbiae* Ulrich.............................. 46
            *P. biconvexus* Ulrich....................... 47
            syn. *P. altimarginatus* Clark.................... 47
    3b. Width greater than length—*P. godoni abbreviatus* Hambach................... 49
        syn. *P. brevis* Ulrich...................... 49
  2b. Deltoids flare; calyx barrel-shaped—*P. rusticus* Hambach........................ 50
1b. Ambulacra concave at top
  2c. Ambulacra shallow, shallow V-shaped at the top—*P. platybasis* Weller................. 50
  2d. Ambulacra deep V-shaped (*see P. sulcatus* group)

## DESCRIPTIONS OF SPECIES OF *PENTREMITES GODONI* GROUP

### Pentremites godoni pinguis Ulrich

(Pl. 3, figs. 1, 2)

*Pentremites pinguis* ULRICH, 1917, Form. Chester Ser. West. Ky., Ky. Geol. Surv., p. 244, Pl. 2, figs. 16?, 17 (Ste. Genevieve, Carrsville and View, Ky.).—WELLER, 1920, Ill. Geol. Surv., Bull. 41, p. 317, Pl. 4, figs. 8, 9, 11 (Shetlerville and Renault formations, Hardin and Randolph cos., Ill.)

Calyx subglobular (*pinguis* means obese), length and width equal, 12–22 mm, greatest width basal; L/W, 1; vault very convex, sides and summit making an even curve; pelvis short, 3 mm, with nearly straight sides; V/P, 3–8; pelvic angle 135°–160°; ambulacra convex in lower part to flat at top, with low, sharp rims; interambulacra slightly concave; 8 transverse ridges in 3 mm; deltoids 4–9 mm less than half the length of the ambulacra, not reaching summit.

Ste. Genevieve limestone to Paint Creek, formation, Ky. and Ill.

This variety is almost identical with *P. godoni*; the main differences are the circular instead of paraboloid outline of the vault and the slightly flatter base. It is even more like *P. godoni abbreviatus*. It differs from *P. platybasis* in having flat instead of concave upper ambulacra. Ulrich's Figure 16 is *P. godoni*.

## Pentremites tuscumbiae Ulrich

(Pl. 3, figs. 3, 4)

*Pentremites tuscumbiae* ULRICH, 1917, Form. Chester Ser. West. Ky., Ky. Geol. Surv., p. 244, Pl. 2 figs. 14, 15 (Ste. Genevieve limestone, Tuscumbia, Ala., Ohara limestone, Carrsville, Ky.)

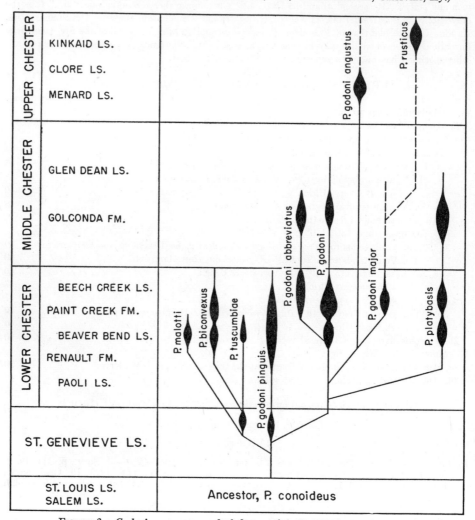

FIGURE 3.—*Geologic occurrence and phylogeny of the* Pentremites godoni *group*

Calyx ovoid, large; up to 24 mm long, av. adult 21 mm; width up to 19 mm, av. adult 18 mm; greatest width at points of ambulacra; L/W, 1.2–1.3; vault with moderately curved sides, av. 15 mm long; pelvis 5–6 mm long, with nearly straight sides; V/P, 3–4, pelvic angle 115°–135°; ambulacra about 3.5 mm wide, moderately convex, with low rims; 7 transverse ridges in 3 mm; interambulacra moderately concave; deltoids 6–7 mm long, ending about 1 mm below summit.

Ste. Genevieve limestone, Tuscumbia, Ala.; Ohara limestone, Carrsville, Ky.; Beaver Bend limestone, Silverville, Ind.

This species is described from Figure 15 in Ulrich, though Figure 14 was designated as the holotype. Since the "slightly crushed" specimen of the original description is evidently Figure 14, the numbers may have been reversed in printing. This species is possibly the same as *P. biconvexus*, but the figures and explanation are inadequate for differentiation. Ulrich presumed that the Ste. Genevieve form should differ from the lower Chester form (*P. biconvexus*), but forms cannot be distinguished only on the basis of slightly different horizons.

TABLE 3.—*Main characters of the* Pentremites godoni *group*

Calices subglobular to conoidal, medium size; vault 3–20 times pelvis; basal angle 110°–170° ambulacra wide, 3–5 mm, slightly convex to flat; ambulacral rims low and narrow; interambulacra slightly concave; deltoids below spiracles. Ste. Genevieve, Chester, doubtfully Lower Pennsylvanian, abundant in the lower and middle Chester. (Synonyms in italics)

| Species | Length/Width | Vault/Pelvis | Pelvic angle | Ambulacra | Inter-ambulacra | Original horizon and locality |
|---|---|---|---|---|---|---|
| P. godoni pinguis | 1 | 3–7 | 135°–160° | convex to flat | slightly concave | Upper Ste. Genevieve; Ky. |
| P. tuscumbiae | 1.2–1.3 | 3–4 | 115°–135° | convex | slightly concave | Ste. Genevieve; Ala. |
| P. biconvexus | 1.3 | 4 | 115°–120° | convex | slightly concave | Lower Chester; Ky. |
| *P. altimarginatus* | 1.2 | 4.5 | 115°–120° | slightly convex | slightly concave | Upper Mississippian; Mont. |
| P. malotti | 1.1 | 2–3 | 105°–115° | convex | nearly flat | Beaver Bend; Ind. |
| P. godoni | 1.1 | 4–8 | 125°–150° | slightly convex | concave | Lower Chester; Ky. |
| *P. planus* | 1 | 9 | 150° | plane | concave | Lower Chester; Tenn. |
| P. godoni abbreviatus | 0.8–1 | 3.8 | 130°–160° | slightly convex | slightly concave | Lower Chester; Ill. |
| *P. brevis* | 1.1 | 3 | 130° | convex | concave | Glen Dean; Ky. |
| P. godoni major | 1.2 | 6–7 | 130°–160° | flat | slightly concave | Lower Chester; Ky. |
| P. godoni angustus | 1.3–1.5 | 8–20 | 130°–160° | slightly convex | slightly concave | Chester; Ark. |
| P. rusticus | 1 | 6–10 | 150°–170° | slightly convex | slightly concave | Lower Pennsylvanian?; Ark. |
| P. platybasis | 0.9–1.1 | 3–6 | 130°–140° | slightly concave | slightly concave | Golconda; Ill. |

## Pentremites biconvexus Ulrich

(Pl. 3, figs. 5–8)

*Pentremites biconvexus* ULRICH, 1917, Form. Chester Ser. West. Ky., Ky. Geol. Surv., p. 247, Pl. 2, figs. 34–36 (Lower Chester, Bowling Green, Ky.).—BUTTS, 1926, Geol. Surv. Ala., Spec. Rept. 14, p. 180, Pl. 59, fig. 2 (Lower Chester, Blount Springs, Ala.)
*Pentremites altimarginatus* CLARK, 1917, Bull. Mus. Comp. Zool., Harvard, vol. 61, no. 9, p. 366, Pl. 1, figs. 11–13 (Mississippian, Old Baldy Mt., Mont.)

Calyx ovoid, 20–30 mm high, 16–24 mm wide, greatest diameter submedian; L/W, 1.2–1.3; vault with very convex sides and narrow summit; length 20–24 mm; pelvis 5–7 mm high, with concave sides; V/P, 3–4; pelvic angle 115°–120°; ambulacra convex in lower part, biconvex on the two sides of the median groove (hence *biconvexus*), flat at top, 4–7 mm wide, with sharp rims (hence *altimarginatus*); 6–7 transverse ridges in 3 mm. interambulacra slightly concave; deltoids up to 11 mm long, less than half the length of the ambulacra, not reaching summit.

Lower Chester, Ala. and Mont., Beaver Bend limestone, Silverville, Ind., Paint Creek formation, St. Clair Co., Ill., Upper Mississippian, Quadrant formation, Old Baldy, Mont.

This species differs from *P. godoni* only in having a longer pelvis and the consequent smaller pelvic angle. Biconvex ambulacra occur also in *P. godoni* and its varieties. *P. biconvexus* and *P. altimarginatus* are identical point by point. Both descriptions were published in 1917, but which one first cannot be determined. Ulrich's specimens are well preserved, and Clark's poorly preserved, so that it is safer to recognize Ulrich's specific name. *P. biconvexus* is an uncommon form, possibly a variety of *P. godoni*. It differs from the variety *major* in the longer base and more conoidal vault. *P. biconvexus* is scarcely distinguishable from *P. tuscumbiae*.

### Pentremites malotti Galloway and Kaska, n. sp.

#### (Pl. 3, figs. 9, 10)

?*Pentremites princetonensis* ULRICH, (part), 1917, Miss. Form. West. Ky., Ky. Geol. Surv., p. 244, Pl. 2, fig. 10 (error for Fig. 13) (Ste. Genevieve limestone, View, Ky.)

Calyx globular, pentagonal in side view as well as in basal view; width nearly equals length; length 17–21 mm, width 15–20 mm; greatest width suprabasal; L/W, 1–1.1; sides and summit of vault nearly straight; pelvis 5–6 mm long; V/P, 2–3; sides of pelvis nearly straight; pelvic angle 105°–115°; ambulacra strongly convex, becoming flat at top, 4 mm wide; rims sharp and narrow; transverse grooves 6 in 3 mm at bottom of ambulacra, 7 in middle and 8 at top; interambulacra area nearly flat, becoming slightly concave between ambulacral points; deltoids 7 mm long, not quite half the length of the ambulacra, points at level of mouth.

Abundant in the lower Chester, Beaver Bend limestone, quarry half a mile east of Silverville, Lawrence Co., Ind.

This species has the smallest pelvic angle of any member of the *P. godoni* group, except for young specimens. It has straighter pelvic sides than *P. tuscumbiae*. It is wider and has straighter vault sides than *P. princetonensis*. The small number of transverse grooves in 3 mm is noticeable but scarcely different from other species with which *P. malotti* occurs.

This species is named in honor of the late Dr. Clyde A. Malott, who collected the specimens.

Holotype: Ind. Univ. Paleo. Coll., No. 5224, Beaver Bend limestone, Silverville, Ind.

### Pentremites godoni (Defrance)

#### (Pl. 3, figs. 11–13; Pl. 11, figs. 20–30; Pl. 13, figs. 9–12)

"An asterial fossil", PARKINSON, 1808, Organ. Rem., vol. 2, p. 235, Pl. 13, figs. 36, 37 (Chester formation, Ky.)
*Encrina Godonii* DEFRANCE, 1819, Dict. Sci. Nat., vol. 14, p. 467 (Named from Parkinson's figures. Specific name is nomenclaturally valid, although the *G* and *ii* are no longer used)
*Encrinites florealis* VON SCHLOTHEIM, 1820, Petrefactenkunde, p. 339 (Named from Parkinson's figures; therefore an exact synonym of *Encrina Godonii* Defrance)
*Pentremite florealis* SAY, 1825, Jour. Acad. Nat. Sci. Philadelphia, 1st ser., vol. 4, p. 295 (Genotype of genus *Pentremites* Say, one of Say's three original species, designated as genotype by Etheridge and Carpenter, 1886, p. 157, but name of species is *P. godoni* (Defrance), not *P. florealis*)
*Pentatremites florealis* GOLDFUSS, 1826, Petre. Germ., Th. 1, p. 150, Pl. 50, fig. 2a–c
*Pentatrematites florealis* ROEMER, 1851, Archiv. f. Naturgesch., Jahrg. 17, Bd. 1, p. 353, Pl. 4, figs. 1–4; Pl. 5, fig. 8
*Pentremites godoni* HALL, 1858, Rept. Geol. Surv. Iowa, vol. 1, pt. 2, p. 692, Pl. 25, fig. 13 (Chester, Ill., Huntsville, Ala.).—WHITE, 1881, 2nd Ann. Rept., Ind. Dept. Stat. and Geol., p. 511, Pl. 7, figs. 10, 11 (Good figures).—ETHERIDGE AND CARPENTER, 1886, Cat. Blastoidea Geol. Dept. British Mus. (Nat. Hist.), p. 157, Pl. 1, fig. 11; Pl. 2, figs. 1–13; Pl. 12, figs. 16, 17; Pl. 16, figs. 19, 22, 23 (Good description and figures).—WELLER, 1898, U. S. Geol. Surv., Bull. 153, p. 414 (Many references).—ROWLEY, 1903, Contrib. Ind. Paleo., vol. 1, pt. 12, p. 121–125, Pl. 36, figs. 18–38 (Chester, Bowling Green, Ky. Variations and abnormalities).—WELLER, 1920, Ill. Geol. Surv., Bull. 41, p. 319, Pl. 4, figs. 31–34, 36, not 35 (*P. godoni major*) nor 47 (*P. godoni abbreviatus*) (Good discussion of nomenclature and structure).—GREGER, 1934, Trans. Acad. Sci. St. Louis, vol. 28, nos. 3, 4 (Most references).—ULRICH, 1917, Miss. Form. West. Ky., Ky. Geol. Surv., p. 254, Pl. 5, fig. 26 (not figs. 17–25, 27–30 *P. princetonensis* Ulrich)

*Pentremites planus* ULRICH, 1917, Miss. Form. West. Ly., Ky. Geol. Surv., Pl. 5, figs. 1–13. (Lower Chester, Cowan, Tenn.; Bowling Green, Ky.; Floraville, Ill.)

Calyx ovate, 10–22 mm long, 9–20 mm wide, full-grown adults typically 18 mm long and 15 mm wide, greatest width near base of ambulacra; L/W, 1.1–1.2; vault paraboloid in outline, with sides moderately curved and summit narrow and nearly flat; pelvis short, 1–3 mm; V/P, 4–8; sides of pelvis slightly concave; pelvic angle of adults 125°–150°, of half-grown specimens 100°; stem facet small, 1.5–2 mm wide; ambulacra convex at the lower ends, flat or even slightly concave near the spiracles, with low but sharp ambulacral rims; 7 to 9 lateral grooves in 3 mm; interambulacra moderately concave; deltoids normally 7 mm long, concave outside, tips about 1 mm below the spiracles.

Lower Chester, Ky., Ala., Tenn., Ind., Ill., Mo., Mont. and Utah, most abundant in Paint Creek formation, also in Beaver Bend limestone. Abundant in the Golconda at Golconda, Ill., rare in the Golconda formation at Grantsburg, Ind., and in the Fayetteville shale, Fayetteville, Ark.; questionably identified by Clark (1917, p. 370) from the Mississippian at Old Baldy and on the divide between Ross Fork and Lincoln Valley, Mont.

This species is one of the most abundant of the genus; it is the genotype of *Pentremites*. It has been considered the same form by all authors except Ulrich. It has a higher vault in *P. godoni major* and *P. godoni angustus* and a shorter pelvis and vault in *P. godoni abbreviatus*.

## Pentremites godoni abbreviatus Hambach

### (Pl. 3, figs. 14–17)

*Pentremites abbreviatus* HAMBACH, 1880, Trans. St. Louis Acad. Sci., vol. 4, p. 155, Pl. B, fig. 3; 1903, vol. 13, p. 36, fig. 12a (Chester, Evansville, Ill.).—ETHERIDGE AND CARPENTER, 1886, Cat. Blastoidea Geol. Dept. British Mus. (Nat. Hist.), p. 160, Pl. 2, fig. 4 (Chester, Ill.)
*Pentremites godoni* WELLER (part), 1920, Ill. Geol. Surv., Bull. 41, Pl. 4, fig. 47 (Paint Creek formation, St. Clair Co., Ill.)
?*Pentremites brevis* ULRICH, *in* BUTTS, 1917, Miss. Form. West. Ky., Ky. Geol. Surv., p. 100, Pl. 24, fig. 6 (Glen Dean limestone, Ky.).—BUTTS, 1926, Ala. Geol. Surv., Spec. Rept. 14, p. 198, Pl. 65, fig. 3 (Bangor limestone, Marshall Co., Ala.) (Not *P. brevis* WELLER, 1920, Pl. 4, figs. 43–46, which is *P. tulipaformis*)

Calyx depressed, shorter than wide, the type 13 mm long and 16 mm wide, greatest width basal; L/W, 0.8–1; vault with moderately curved sides and flat summit; pelvis short, 1–4 mm, with straight to concave sides; V/P, 3–8; pelvic angle 130°–160°; ambulacra convex at lower ends, flat at upper ends, 3–4 mm wide, with low, sharp rims; 9 lateral grooves in 3 mm; interambulacra moderately concave; deltoids 6–7 mm long, reaching the summit.

Also occurs sparingly in the Golconda formation at Golconda, Ill.

This variety differs from *P. godoni* only in being shorter and generally having a flatter base. It is even more compressed dorsoventrally than *P. godoni pinguis*. The ambulacra are nearly flat, as in *P. godoni*, not strongly convex, as in *P. conoideus amplus*. There is about one specimen of the variety to 10 specimens of the species. The U. S. National Museum specimens of *P. brevis* Ulrich are identical with *P. abbreviatus* Hambach.

## Pentremites godoni major Etheridge and Carpenter

### (Pl. 3, figs. 20–22)

*Pentremites godoni* var. *major*, ETHERIDGE AND CARPENTER, 1886, Cat. Blastoidea Geol. Dept. British Mus. (Nat. Hist.), p. 160, Pl. 2, fig. 2 (Chester, Franklin Co., Tenn.).—ULRICH, 1917, Form. Chester Ser. West. Ky., Ky. Geol. Surv., Pl. 5, fig. 15. (Lower Chester, Cowan, Tenn.).—WELLER, 1920, Ill. Geol. Surv. Bull. 41, Pl. 4, fig. 35 (Paint Creek formation, St. Clair Co., Ill.)

Calyx barrel-shaped, up to 28 mm long and 22 mm wide, greatest width submedian; L/W, 1.2; vault with moderately curved sides and flat summit, 20–23 mm high; pelvis short, about 3 mm, with nearly straight sides; V/P, 6–7; pelvic angle 130°–160°; ambulacra convex at the lower ends, flat at the upper ends, 5 mm wide between deltoids, with sharp rims 0.5 mm high; 7–9 transverse grooves in 3 mm; interambulacra moderately concave; deltoids less than half length of ambulacra, nearly reaching spiracles

Lower Chester, Tenn., Warren Co., Ky., and Paint Creek formation, Floraville, Ill.

This variety is larger and longer than *P. godoni*, with less-curved sides and flatter base. It is about one-tenth as abundant as the species and may be gerontic individuals of *P. godoni*.

### Pentremites godoni angustus Hambach

(Pl. 3, fig. 23, 24; Pl. 13, fig. 1)

*Pentremites angustus* HAMBACH, 1903, Trans. St. Louis Acad. Sci., vol. 13, p. 53, Fig. 14 (Chester limestone, Washington Co., Ark.).—MATHER, 1916, Bull. Sci. Lab. Denison Univ., vol. 18, p. 100, pl. 3, figs. 10–12 (Lower Pennsylvanian?, Brentwood limestone, 2 miles northwest of Brentwood, Ark.)

Calyx oblong, up to 22 mm high and 16 mm wide, the greatest width basal or suprabasal; L/W 1.3–1.5; vault high, 18–21 mm, with sloping, moderately curved sides and convex summit; pelvis short, 1–2 mm, with slightly concave sides; V/P, 8–20; pelvic angle 130°–160°; ambulacra slightly convex, with low, sharp rims; 7 transverse ridges in 3 mm; interambulacra slightly concave; deltoids 8–10 mm long, not reaching summit.

Chester limestone, Ark., Lower Pennsylvanian?, Brentwood and Hale formations, southern Ark. and northeast Okla., and Paint Creek formation, St. Clair Co., Ill.

This form seems to be a variety of *P. godoni*, analogous to the variety *P. conoideus perlongus*. Figure 13a of Plate 3 in Mather, 1916, seems to be exactly *P. godoni*. The Brentwood age of *P. godoni angustus* is anomalous, supposedly Lower Pennsylvanian, but probably lower Chester.

### Pentremites rusticus Hambach

(Pl. 3, figs. 18, 19)

*Pentremites rusticus* HAMBACH, 1903, Trans. St. Louis Acad. Sci., vol. 13, p. 54, Fig. 15 (Chester limestone, Washington Co., Ark.).—MATHER, 1916, Bull. Sci. Lab. Denison Univ., vol. 18, p. 101, Pl. 3, figs. 3–6 (Brentwood limestone, northeast Okla.)

Calyx keg-shaped, length and width subequal, up to 23 mm, greatest width median; L/W, 1; vault with slightly curved sides and truncate summit; pelvis short, 1–2 mm, with nearly straight sides; V/P, 6–10; pelvic angle 150°–170°; ambulacra slightly convex 5 mm wide at top, with shallow median groove, and surrounded by low narrow rim which is progressively higher toward top; transverse grooves 8–10 in 3 mm; interambulacra slightly concave below, narrowly concave in deltoids; deltoids flaring, up to 9 mm long, but less than half the length of the ambulacra, not reaching summit; basals stellate. Young specimens tend to be globular and much like young of *P. godoni*.

Morrow group, Lower Pennsylvanian, possibly Upper Mississippian, Arkansas and Oklahoma. The authors have many specimens of this species from Arkansas, locality and horizon unknown.

This is one of the best-characterized and least variable species of the genus. The flaring deltoids are similar to those of the *P. sulcatus* group, but the convex ambulacra and the lower deltoids differentiate it from all species of that group. Hambach says (p. 54), "It belongs in the Chester limestone, but has been so far known only from Washington County in Arkansas." Mather did not find it in Arkansas; he considers its age to be Lower Pennsylvanian, which may be questioned.

### Pentremites platybasis Weller

(Pl. 3, figs. 25–29)

*Pentremites platybasis* WELLER, 1920, Ill. Geol. Surv., Bull. 41, p. 355, Pl. 4, figs. 37–42 (Lower Okaw = Golconda limestone, Randolph Co., Ill.)

Calyx globular, up to 14 mm in length and width; greatest width submedian; L/W, 0.9–1.1; sides of vault slightly curved, with convex summit; pelvis 2–3 mm long, with very slightly concave sides; V/P, 3–6; pelvic angle 130°–140°; ambulacra flat below and slightly concave at summit, 3–4 mm wide; lateral grooves 9–12 in 3 mm; interambulacra slightly concave, with narrow rims; deltoids short, 2–4 mm, not reaching summit.

Lower Chester, Beaver Bend limestone, Silverville, Ind., Paint Creek formation, St. Clair Co., Ill.; middle Chester, Golconda limestone, Illinois; Shoals and Grantsburg, Ind.

This species is shaped like *P. godoni* from which it was derived but differs in the concave upper ambulacra and in having more lateral grooves in 3 mm. It shows the beginning of the concave ambulacra of the *P. sulcatus* group of the middle and upper Chester. Young specimens have higher pelves and smaller pelvic angles, 90°–120°, than the adults.

## KEY TO SPECIES OF *PENTREMITES PYRIFORMIS* GROUP

Page

1a. Basal angle 100°–115°
  2a. Vault evenly rounded; V/P, 3—*P. pulchellus* Ulrich............................... 51
  2b. Vault a truncate pyramid; V/P, 2
    3a. Calyx small; L/W, 1.1—*P. princetonensis* Ulrich.............................. 52
      syn. *P. divergens* Clark........................................ 54
    3b. Calyx large; L/W, 1.3—*P. abruptus* Ulrich................................. 54
1b. Basal angle, 80°–95°
  2c. Vault-pelvis ratio, 2
    3c. Length over width, 1.3–1.5—*P. symmetricus* Hall.............................. 55
      syn. *P. decipiens* Ulrich....................................... 55
      syn. *P. decipiens decurtatus* Ulrich................................ 55
      syn. *P. saxiomontanus* Clark..................................... 55
    3d. Length over width, 1.6—*P. altus* Rowley................................ 55
  2d. Vault-pelvis ratio, 1.5—*P. welleri* Ulrich................................. 56
      syn. *P. lyoni* Ulrich......................................... 56
      syn. *P. arctibrachiatus* Ulrich................................... 56
      syn. *P. arctibrachiatus huntsvillensis* Ulrich.......................... 56
      syn. *P. pediculatus* Ulrich...................................... 56
1c. Basal angle, 60°–75°
  2e. Vault and pelvis subequal in length; V/P, 0.9–1.2
    3e. Vault evenly rounded
      4a. Pelvic angle nearer 60°; vault pyramidal—*P. pyriformis* Say..................... 56
      4b. Pelvic angle nearer 70°; vault bulbous—*P. patei* Ulrich....................... 57
    3f. Vault pyramidal; pelvic angle, 70°–75°—*P. pyramidatus* Ulrich.................... 57
  2f. Vault much longer than pelvis; V/P, 1.6–1.8
    3g. Calyx broad; L/W, 1.3—*P. gemmiformis* Hambach............................ 58
    3h. Calyx very elongate; L/W, 1.7—*P. buttsi* Ulrich............................ 58
1d. Basal angle 35°–60°
  2g. Vault longer than pelvis—*P. kirki* Hambach................................ 58
      syn. *P. lyoni gracilens* Ulrich.................................... 58
  2h. Vault shorter than pelvis
    3i. Vault half the length of pelvis; V/P, 2—*P. turbinatus* Hambach.................... 59
    3j. Vault more than half length of pelvis
      4c. Length over width less than 2; pelvic angle 50°–60°
        5a. Length over width, 1.3—*P. speciosus* Rowley...................... 59
        5b. Length over width, 1.4–1.5—*P. girtyi* Ulrich...................... 59
        5c. Length over width, 1.6–1.7—*P. okawensis* Weller.................... 60
      4d. Length twice the width; pelvic angle 35°–40°—*P. clavatus* Hambach.............. 60

## DESCRIPTION OF SPECIES OF *PENTREMITES PYRIFORMIS* GROUP

### Pentremites pulchellus Ulrich

(Pl. 4, figs. 1–4)

*Pentremites pulchellus* ULRICH, 1917, Form. Chester Ser. West. Ky., Ky. Geol. Surv., p. 242, Pl. 2, figs. 1–7 (Ste. Genevieve, Fredonia oolite, 2 miles southeast of Fredonia, Ky.).—WELLER, 1920, Ill. Geol. Surv., Bull. 41, p. 316, Pl. 4, 28–30 (Fredonia member, Princeton, Ky., Rosiclare, Ill.; Shetlerville formation, Hardin Co., Ill.).—BUTTS, 1926, Geol. Surv. Ala., Spec. Rep. 14, p. 181, Pl. 59, fig. 33 (Ste. Genevieve limestone, Huntsville, Ala.)

Calyx subovate or olivoid, typically 13 mm long and 10 mm wide, greatest width suprabasal; L/W, 1.3–1.5; vault parabolic in outline; pelvis 3 mm long, with nearly straight sides; V/P, 3–3.3; basal angle 105°–110°; ambulacra moderately convex and moderately wide, 3 mm at tips of deltoids,

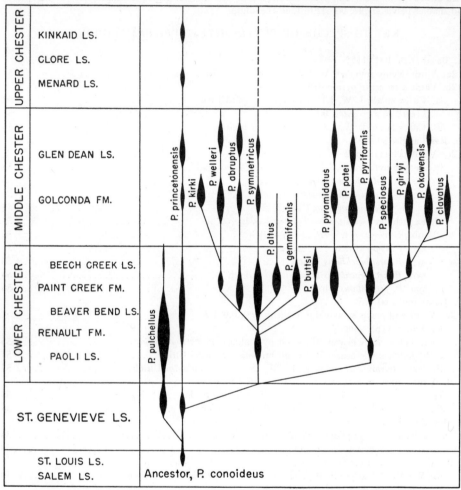

FIGURE 4.—*Geologic occurrence and phylogeny of the* Pentremites pyriformis *group*

with very low and narrow rims; lateral grooves, 8–9 in 3 mm; interambulacra slightly concave; deltoids 3 mm long, not reaching summit.

Lower Ste. Genevieve to Paint Creek, most abundant in the Downey's Bluff member of the Renault formation; Ky., Ala., Ill.

This species is intermediate in structure between the typical members of the *P. conoideus* and the *P. pyriformis* groups of species. It has a longer pelvis than *P. conoideus perlongus* and *P. godoni angustus*, with which it might be confused. The basal angle is larger than in any other species of the *P. pyriformis* group.

### Pentremites princetonensis Ulrich

(Pl. 4, figs. 5–8)

*Pentremites princetonensis* ULRICH, 1917, Form. Chester Ser. West Ky., Ky. Geol. Surv., p. 243–244, Pl. 2, figs. 8–13 (Fredonia oölite, Ste. Genevieve limestone, Princeton, Marion, Carrsville,

TABLE 4.—*Main characters of the* Pentremites pyriformis *group*

Calices ovate, pyriform, bipyramidal to clavate, small to large; vault pyramidal to paraboloid, 0.5–3 times pelvis; pelvis pyramidal, low to high; pelvic angle 35°–115°; ambulacra slightly convex to flat, with low rims; deltoids end below spiracles. Ste. Genevieve to Upper Chester. (Synonyms in italics)

| Species | Length/Width | Vault/Pelvis | Pelvic angle | Ambulacra | Inter-ambulacra | Original horizon and locality |
|---|---|---|---|---|---|---|
| P. pulchellus | 1.3–1.5 | 3 | 105°–110° | convex | slightly concave | Ste. Genevieve; Ky. |
| P. princetonensis | 1.1–1.2 | 2 | 100° | slightly convex | slightly concave | Ste. Genevieve; Ky. |
| *P. divergens* | 1.1 | 2 | 100°? | flat | nearly flat | Upper Mississippian; Mont. |
| P. abruptus | 1.2–1.3 | 1.8–2 | 90°–100° | slightly convex | slightly concave | Lower Chester; Tenn. |
| P. symmetricus | 1.3–1.5 | 2.2–3 | 80°–90° | slightly convex | nearly flat | Lower Chester; Ky. |
| *P. decipiens* | 1.5 | 1.8 | 75° | flat | slightly concave | Lower Chester; Ky. |
| *P. decipiens decurtatus* | 1.4 | 2.1 | 90° | flat | nearly flat | Lower Chester; Ky. |
| *P. saxiomontanus* | 1.4 | 2.1 | 95° | flat | flat | Upper Mississippian; Mont. |
| P. altus | 1.6 | 2–3 | 85°–95° | slightly convex | slightly concave | Lower Chester; Tenn. |
| P. welleri | 1.4 | 1.5 | 85° | flat | slightly concave | Lower Chester; Ala. |
| *P. lyoni* | 1.5 | 1.8 | 80° | flat | slightly concave | Golconda; Ky. |
| *P. arctibrachiatus* | 1.3 | 1.6 | 90° | flat | nearly flat | Lower Chester; Ky. |
| *P. arctibrachiatus huntsvillensis* | 1.2–1.4 | 1–1.5 | 85°–110° | flat | concave | Lower Chester; Ala. |
| *P. abruptus* var. | 1.3 | 1.4 | 90° | flat | nearly flat | Lower Chester; Ala. |
| *P. pediculatus* | 1.5 | 1.1 | 80°–90° | slightly convex | slightly concave | Lower Chester; Ky. |
| P. pyriformis | 1.5 | 0.9 | 60°–65° | flat to slightly convex | flat | Lower & Middle Chester; Ala. |
| P. patei | 1.3 | 1.1–1.2 | 65°–70° | flat | flat | Lower Chester; Ky. |
| P. pyramidatus | 1.2 | 1.2 | 70°–75° | slightly convex | flat | Middle Chester; Ala. |
| P. gemmiformis | 1.3 | 1.8 | 70°–75° | flat | nearly flat | Lower Chester; Ill. |
| P. buttsi | 1.7 | 1.7–2 | 75°–90° | flat | flat | Lower Chester; Ill. |
| P. kirki | 2.1 | 1.6 | 50°–60° | nearly flat | concave? | Burlington ?? Loc.? |
| *P. lyoni gracilens* | 2 | 1.4 | 50°–60° | nearly flat | flat | Lower Chester; Ill. |

TABLE 4.—*Continued*

| Species | Length / Width | Vault / Pelvis | Pelvic angle | Ambulacra | Inter- ambulacra | Original horizon and locality |
|---|---|---|---|---|---|---|
| P. turbinatus | 1.2 | 0.5 | 50° | flat | flat | Middle Chester; Ill. |
| P. speciosus | 1.3 | 0.7–0.8 | 55°–60° | flat | flat | Middle Chester; Ky. |
| P. girtyi | 1.5 | 0.7 | 50°–60° | flat | flat, concave at top | Lower Chester; Ky. |
| P. okawensis | 1.6 | 0.7 | 50°–55° | nearly flat | flat | Golconda; Ill. |
| P. clavatus | 2 | 0.7 | 35°–40° | slightly convex | flat | Golconda; Ill. |

and View, Ky., also Renault and Paint Creek, Ill.).—WELLER, 1920, Ill. Geol. Surv., Bull. 41, p. 314–316, Pl. 2, figs. 1–7 (St. Louis limestone, St. Louis, Mo. and Ste. Genevieve Co., Mo.; Ste. Genevieve limestone, Monroe Co., Ill., Ste. Genevieve Co., Mo.; Fredonia limestone, Princeton, Ky., Rosiclare, Ill.; Shetlerville formation, Hardin Co., Ill., and elsewhere; Renault formation, Johnson Co., Ill., and elsewhere)

*Pentremites divergens* CLARK, 1917, Bull. Comp. Zool., vol. 61, p. 365, Pl. 1, figs. 7–10 (Quadrant formation, Old Baldy, Virginia City, Mont.)

Calyx subglobular, small; av. adult 12 mm long and 10 mm wide; greatest width suprabasal; L/W, 1.1–1.2; vault a truncate pyramid; pelvis 4 mm long, obconical; V/P, 2; basal angle 100°; ambulacra slightly convex, not expanding widely at summit, 3 mm wide at tips of deltoids; 9–10 transverse grooves in 3 mm; ambulacral margins low, interambulacra slightly concave; deltoids 3–4 mm long, not extending to summit.

St. Louis to Upper Chester, Ky., Ill., Mo., Ala., most common in the lower Chester. Rare in the late Chester, Kinkaid limestone, 8 miles northeast of Marion, Ky.; Upper Mississippian, Quadrant formation, Old Baldy, Mont.

*Pentremites princetonensis* is thoroughly discussed by Weller (1920 p. 315.) He concludes that Ulrich had confused the three species *P. princetonensis*, *P. pinguis*, and *P. pulchellus*. *P. pinguis* is a variety of *P. godoni*. The other two species are distinguishable by the greater V/P ratio in *P. pulchellus*. *P. divergens* is too poorly preserved for recognition, but does not differ from *P. princetonensis*. All the species of *Pentremites* discussed by Clark (1917, p. 361–371) belong to the groups with convex or flat ambulacra; none belong to the *P. sulcatus* group with concave ambulacra; hence the age of the fauna ranges from Meramecian to lower Chester.

### Pentremites abruptus Ulrich

(Pl. 4, figs. 9–11)

*Pentremites abruptus* ULRICH, 1917, Form. Chester Ser. West. Ky., Ky. Geol. Surv., p. 257, Pl. 6, figs. 10, 11, 14, fig. 11 holotype (Lower Chester, Cowan, Tenn.)

Calyx ovoid, large, up to 32 mm long, av. adult 27 mm; width up to 22 mm, av. adult 21 mm; greatest width at points of ambulacra; L/W, 1.2 = 1.3; vault with nearly straight sides and truncate summit; pelvis with nearly straight sides; V/P, 1.8–2, pelvic angle 90°–110°, av. adult 100; ambulacra broad, about 6 mm at tips of deltoids, slightly convex, with low rims; 6–7 transverse ridges in 3 mm interambulacra gently concave; deltoids 5–7 mm long, 6 mm in typical adult, ending 1 mm below summit.

Lower Chester: Monte Sano limestone, Huntsville, Ala., Cowan and Sneedsville, Tenn.; Paint Creek formation, Floraville, Ill. Middle Chester: Glen Dean limestone, Cloverport, Ky. and Leopold, Perry Co., Ind. Chester Series, Breckinridge Co., Ky.

The typical form of *P. abruptus* has convex ambulacral areas, but Ulrich mentions that some Cowan specimens are more slender and nearly half belong to a variety which possess flat to concave ambulacra. This species is very similar to *P. symmetricus*, but is not so slender, with a smaller L/W ratio and flat summit.

## Pentremites symmetricus Hall

(Pl. 4, figs. 12–18)

*Pentremites symmetricus* HALL, 1858, Rept. Geol. Surv. Iowa, vol. 1, pt. 2, p. 694, Pl. 25, fig. 14 (Chester limestone, Ky.).—ULRICH, 1917, Form. Chester Ser. West. Ky., Ky. Geol. Surv., p. 260, Pl. 7, fig. 9. (Paint Creek, 5 miles northeast of Waterloo, Ill.).—WELLER, 1920, Ill. Geol. Surv. Bull. 41, p. 324, Pl. 4, figs. 26–27 (Renault and Paint Creek limestones, St. Clair and Monroe Co., Ill.)
*Pentremites decipiens* ULRICH, 1917, Form. Chester Ser. West. Ky., Ky. Geol. Surv., p. 254, Pl. 5, figs. 31–33 (Lower Chester, Cowan, Tenn.)
*Pentremites decipiens decurtatus* ULRICH, 1917, Form. Chester Ser. West. Ky., Ky. Geol. Surv., p. 255, Pl. 5, figs. 34–35 (Lower Chester, Breckenridge Co., Ky. and Cowan, Tenn.)
*Pentremites saxiomontanus* CLARK, 1917, Bull. Mus. Comp. Zool., vol. 61, no. 9, p. 363, Pl. 1, figs. 1–6, 14 (Quadrant formation, Old Baldy near Virginia City, Mont.)

Calyx ovate, up to 32 mm long, av. adult 20 mm; width up to 21 mm, av. adult 15 mm; greatest width at base of ambulacra; L/W, 1.4–1.5; vault with nearly straight lateral profiles, up to 23 mm long, av. 15 mm; summit flat; pelvis 7–8 mm long with nearly straight edges; V/P, 2–3.3, av. 2.3; pelvic angle 80°–90°; ambulacra 3–4.5 mm wide, convex at point, flat at top, with low, narrow rims; 7–8 transverse ridges in 3 mm; interambulacra nearly flat; deltoids 5–8 mm long, generally ending about 1 mm below summit.

Chester limestone, Ky.; Paint Creek formation, Randolph and St. Clair cos., Ill., Renault formation, Floraville, Ill., St. Clair and Monroe cos., Ill.; Beaver Bend limestone, Stanford, Ind.; Glen Dean, Cloverport, Ky.; Upper Mississippian, Quadrant formation, Squaw Creek and Old Baldy, Mont.

Ulrich (1917, p. 260) proposed that one of his specimens (Pl. 7, fig. 9) be used to "re-establish" the species, since Hall's original specimen "seems to be lost". Such action is nomenclaturally illegal and unnecessary, since Hall's figure and description are adequate for recognition and to maintain the species. The specimens illustrated by Ulrich (1917, Pl. 7, figs. 8, 10) are unusually large and may be gerontic individuals, or even belong to *P. abruptus* or *P. welleri. Pentremites symmetricus* is distinguished from *P. welleri* by its shorter pelvis and wider summit.

## Pentremites altus Rowley

(Pl. 4, figs. 19, 20)

*Pentremites altus* ROWLEY, *in* GREENE, 1901, Contrib. to Ind. Paleo., vol. 1, no. 8, p. 64, Pl. 23, fig. 1, not 2, 3, which is *P. buttsi* Ulrich (Holotype, Fig. 1) (Chester limestone Newman's Ridge, Tenn., Bowling Green, Ky.)

Calyx biconical, very large, up to 37 mm long, av. adult 29 mm; width up to 23 mm, av. adult 19 mm; greatest width submedian; L/W, 1.5– 1.6; vault with slightly convex sides, up to 25 mm, av. adult 20 mm; pelvis 10–13 mm long with slightly inflated basals; V/P, 2–3; pelvic angle 85°–95°; ambulacra 4–5 mm wide, slightly convex, becoming flat between deltoid tips, most convex on the two sides of the median groove; rims low and narrow; about 6 transverse ridges in 3 mm, ridges meet at angles at median groove, becoming transverse at top of ambulacrum; interambulacra slightly concave; deltoids 5–10 mm long, ending below summit.

Chester limestone, Newman's Ridge, Tenn., Bowling Green, Ky.; Kinkaid limestone, 2 miles west of Robbs, Ill., collected by Dr. R. C. Gutschick, University of Notre Dame.

*Pentremites altus* is closely related to *P. symmetricus* Hall. The type specimen is distorted; three specimens studied are much compressed laterally. The two species differ mainly by the large length-width ratio of *P. altus*. The lower part of the "X" on the type figure is the trace of growth lines between the deltoid and the radials. It occurs in other species of *Pentremites*, especially specimens which are well preserved but is not a specific character.

## Pentremites welleri Ulrich

(Pl. 4, figs. 21–31)

*Pentremites welleri* ULRICH, 1917, Form. Chester Ser. West. Ky., Ky. Geol. Surv., p. 258, Pl. 6, figs. 15, 18, 23, 24, 26 (Lower Chester limestone, Huntsville, Ala., Bowling Green, Ky.; Renault formation, Floraville, Ill.; Paint Creek formation, Monroe Co., Ill.).—BUTTS, 1941, Va. Geol. Surv. Bull. 52, p. 249, Pl. 132, fig. 14 (Lower Chster, Mercer Co., W. Va.)
*Pentremites lyoni* ULRICH, 1917, Form. Chester Ser. West. Ky., Ky. Geol. Surv., p. 262, Pl. 7, figs. 27–28 (not 29 = *P. pyramidatus*). (Golconda formation, Marion, Ky.; Glen Dean limestone, Sloans Valley, Ky.; Ala., Ga., Tenn., Ill.; Okaw limestone, Randolph Co., Ind.).—HAAS, 1945, Am. Mus. Novitates, no. 1289, p. 4, Figs. 29–31, 35. (Glen Dean, Crane, Martin Co., Ind.)
*Pentremites arctibrachiatus* ULRICH, 1917, Form. Chester Ser. West. Ky., Ky. Geol. Surv., p. 248, Pl. 2, figs. 37–40 (Lower Chester, limestone, Huntsville, Ala.; Renault formation, Floraville, Ill.)
*Pentremites arctibrachiatus huntsvillensis* ULRICH, 1917, Form. Chester Ser. West. Ky., Ky. Geol. Surv. p. 248, Pl. 2, figs. 41–42 (Lower Chester, Huntsville, Ala.)
*Pentremites abruptus* var. ULRICH, 1917, Form. Chester Ser. West Ky., Ky. Geol. Surv., p. 248, Pl. 2, fig. 43 (Lower Chester limestone, Huntsville, Ala.)
*Pentremites pediculatus* ULRICH, 1917, Form. Chester Ser. West. Ky., Ky. Geol. Surv., p. 248, Pl. 2, fig. 44 (Lower Chester, Bowling Green, Ky.)

Calyx ovoid, large; length up to 35 mm; greatest width submedian, up to 25 mm; L/W, 1.3–1.5; vault paraboloid, with slightly convex summit; pelvis pyramidal with slightly concave edges; V/P; 1.5; pelvic angle 85°–95°; ambulacra flat with low, narrow margins; 7 to 8 lateral grooves in 3 mm; interambulacra slightly concave; deltoids 5–10 mm long, usually extending to near summit.

Lower Chester, Ala., Ky., Ill.; Golconda formation, Marion, Ky., and Crawford Co., Ind.; Glen Dean limestone, Ky., Ala., Ga., Tenn., Ill.

Ulrich supposed that *P. lyoni* was derived from *P. welleri* and distinguished them on the basis of more numerous "poral pieces" in the former; they are identical in all other respects. Since the number of poral pieces or lateral ridges is a function of ontogenetic age, not of species, the two are synonyms. *P. welleri* is very similar to *P. abruptus* but is distinguished by its lesser basal angle and more rounded summit. *P. arctibrachiatus* and its varieties differ in no measurable way, excepting size, from *P. welleri*; the types of both species and the varieties of *P. arctibrachiatus* are from the same locality and horizon, lower Chester, Huntsville, Ala. The name *P. welleri* is chosen because it represents the adult form and is therefore better characterized than young specimens, to which the names *P. arctibrachiatus*, *P. arctibrachiatus huntsvillensis*, *P. abruptus* var., and *P. pediculatus* were given by Ulrich. The long stem of *P. pediculatus* means only that one or two columnals were still attached.

## Pentremites pyriformis Say

(Pl. 4, figs. 32–37; Pl. 13, figs. 2, 3)

*Pentremites pyriformis* SAY, 1825, Jour. Acad. Nat. Sci., Philadelphia, (1) vol. 4, p. 294 (Kentucky).—TROOST, 1835, Trans., Penna. Geol. Soc., vol. 1, pt. 2, p. 228, Pl. 10, fig. 8 (Mississippian, Sparta, Tenn., Monte Sano, Ala., Ky.).—HALL, 1858, Rept. Geol. Surv. Iowa, vol. 1, pt. 2, p. 693, Pl. 25, fig. 16 (Chester limestone, Chester, Ill., Huntsville, Ala.).—MILLER, 1889, N. A. Geol. and Paleo., p. 268, Figs. 386, 387 (Chester limestone).—ULRICH, 1905, U. S. Geol. Surv., Prof. Pap. 36, p. 57, Pl. 6, figs. 9–12 (Lower Chester, Ky.).—ULRICH, 1917, Form. Chester Ser. West. Ky. Geol. Surv., p. 257, Pl. 6, figs. 3–6, 8, 9 (not 1, 2, 7) (Lower Chester, Huntsville, Ala., Bowling Green, Ky.)
*Pentremites buttsi* WELLER (not *P. buttsi* ULRICH, 1917), 1920, Ill. Geol. Surv. Bull. 41, p. 323, Pl. 4, figs. 16–20 (Renault limestone, Ill.)

Calyx pyriform, size medium to large, from 20–37 mm long and 14–26 mm wide; greatest width median; L/W, 1.4–1.5; vault parabolic, with rounded summit; pelvis pyramidal with nearly straight edges; V/P, 0.9–1; pelvic angle 60°–65°; ambulacra flat to slightly convex, without or with low, narrow rims; 7 to 8 lateral grooves in 3 mm; interambulacra flat to slightly concave; deltoids short, 2–4 mm long, depending on size of specimen, not reaching summit.

Lower Chester, Renault limestone, Monroe, St. Clair, Union and Hardin cos., Ill.; Paint Creek formation, St. Clair Co., Ill.; Cowan, Tenn.; middle Chester, common in the Golconda formation, Ala., Ill., Ky., Martin, Crawford and Perry counties, Ind.

*Pentremites pyriformis* is more slender than *P. pyramidatus*, and the vault does not converge so rapidly. Ulrich (1917, p. 261) recognized that *P. pyriformis* was probably the species from which *P. girtyi* was derived. The many neanic specimens shown in Ulrich's figures of *P. pyriformis, P. patei*, and *P. girtyi* are similar and of no stratigraphic value. Weller (1920, p. 324) realized that the form he called *P. buttsi* was really *P. pyriformis*, for he stated, "This is the earliest representative of the elongate type of pentremite that has been referred commonly to *P. pyriformis*."

The characters of *P. pyriformis* have been based on the description and figures by Hall, 1858, by most authors, and their forms consistently approximate Hall's form. The vault and pelvis should be of about the same length and shape. Forms with shorter pelves are now placed in *P. welleri, P. abruptus, P. symmetricus, P. altus*, and other species. It is best not to disturb the usage of nearly a century, and Hall's identification should stand.

Plesiotypes: Ind. Univ. Paleo. Coll., No. 5238, Glen Dean, 1 mile East of Cloverport, Ky.; No. 5239, Golconda limestone, half a mile north of Grantsburg, Ind.

### Pentremites patei Ulrich

### (Pl. 5, figs. 1, 2)

*Pentremites patei* ULRICH, 1917, Form. Chester Ser. West. Ky., Ky. Geol. Surv., p. 261, Pl. 7, figs. 17–22 (Lower Chester and Glen Dean formations, Breckinridge, Todd, and Caldwell cos., Ky., Monroe Co., Ill., Huntsville, Ala.).—BUTTS, 1941, Va. Geol. Surv. Bull. 52, p. 250, Pl. 132, fig. 23 (Mercer co., Va.)

Calyx short pyriform, medium size, adults about 15–20 mm long and 12–17 mm wide; greatest width submedian; L/W, 1.1–1.2; vault parabaloid; pelvis pyramidal, edges nearly straight; V/P, 1.1–1.2; pelvic angle 65°–70°; ambulacra nearly flat with low narrow rims; 8 transverse grooves in 3 mm; interambulacra nearly flat; deltoids 2–4 mm long, not quite reaching summit.

Common in the Golconda and Glen Dean formations, Kentucky, Illinois, Alabama, and Perry and Crawford cos., Indiana.

This species is distinguished from *P. girtyi* by its greater basal angle and greater width. It is intermediate between *P. pyriformis* and *P. pyramidatus* and is scarcely distinguishable from either.

Plesiotype: Ind. Univ. Paleo. Coll., No. 5240, Golconda limestone, half a mile north of Grantsburg, Ind.

### Pentremites pyramidatus Ulrich

### (Pl. 5, figs. 3, 4; Pl. 13, fig. 7)

*Pentremites pyramidatus* ULRICH, 1905, U. S. Geol. Surv., P. P. 36, p. 64, Pl. 7, figs. 12–14 (Middle and Upper Chester, W. Ky.).—WELLER, 1920, Ill. Geol. Surv., Bull. 41, p. 325, Pl. 4, figs. 21–24 (Paint Creek formation, Monroe and Johnson cos., Ill.)

Calyx bipyramidal, medium to large; 15–24 mm long and 10–20 mm wide; greatest width median; L/W, 1.2; vault pyramidal with slightly curved and strongly converging sides; pelvis obpyramidal, with straight edges and flat sides; V/P, 1.2; pelvic angle 70°–75°; ambulacra slightly convex with slightly upturned, narrow rims; 6–8 transverse ridges in 3 mm; interambulacra flat; deltoids up to 6 mm in adults, not quite attaining summit.

Lower Chester, Paint Creek formation, Ill., common in the Beech Creek limestone, Ind.; middle Chester, Golconda formation, Ky., Glen Dean limestone, Ky., Ind., Ala.; late Chester, Kinkaid limestone, 8 miles northeast of Marion, Ky.

This species is characterized by the nearly equal and similar vault and pelvis. Ulrich's Figure 13 is here selected as lectotype; it was also considered the typical form by Weller (1920, p. 326). Ulrich did not designate a type specimen; it is irrelevant that a specimen in the U. S. National Museum has "Holotype" lettered on the label.

Plesiotype: Ind. Univ. Paleo. Coll., No. 5241, Golconda limestone, 1–2 miles south of Herod, Ill.

## Pentremites gemmiformis Hambach

(Pl. 5, figs. 5, 6)

*Pentremites gemmiformis* HAMBACH, 1884, Trans. St. Louis Acad. Sci., vol. 4, p. 553, Pl. D., fig. 5 (Lower Chester, probably Paint Creek, Randolph Co., Ill.)—ULRICH, 1917, Form. Chester Ser. West. Ky., Ky. Geol. Surv., p. 260, Pl. 7, figs. 1, 2, 6, not figs. 3–5, 7 (pelvis too long).— WELLER, 1920, Ill. Geol. Surv., Bull. 41, p. 326, Pl. 10, figs. 9, 10 (Paint Creek, St. Clair, and Randolph cos., Ill.)

Calyx ovoid, small, up to 21 mm long (including stem) and up to 13 mm wide; greatest width submedian; L/W 1.3 (not including bulbous stem); vault broad paraboloid, with slightly convex sides and summit; pelvis pyramidal, with straight edges; V/P, 1.6–1.8 (not including bulbous stem); pelvic angle 70°–75°; ambulacra flat, with low, narrow rims; 7 transverse grooves in 3 mm; interambulacra nearly flat; deltoids short, 2–3 mm, not reaching summit.

Lower Chester, Paint Creek, Ill., Ky.

This form is like the neanic stage of *P. welleri*, differing only in the bulbous stem, which is not a specific character, but only a columnal not yet severed from the calyx. Such a beadlike but narrow columnal occurs on many specimens of many species. The columnal is not unusually wide except in the type specimen and Ulrich's Figure 2, in which the width is probably only an individual, or possibly a pathologic, character. The name *P. gemmiformis* must stand, having priority, even if it is the young of some other form.

## Pentremites buttsi Ulrich

(Pl. 5, figs. 7–9)

*Pentremites altus* ? ROWLEY, *in* Greene, 1901, Contrib. to Ind. Paleo., vol. 1, no. 8, p. 64, Pl. 23, figs. 2–3 (Chester limestone, Bowling Green, Ky.)
*Pentremites buttsi* ULRICH, 1917, Form. Chester Ser. West. Ky., Ky. Geol. Surv., p. 245, Pl. 2, figs. 18–19 (Lower Chester, Downeys Bluff, Rosiclare, Ill., Hardin Co., Ill.)
Not *Pentremites buttsi* WELLER, 1920, Ill. Geol. Surv., Bull. 41, p. 323, Pl. 4, figs. 16–20 = *P. pyriformis*

Calyx elongate ovoid; up to 24 mm long, av. adult 18 mm; width up to 14 mm, av. adult 12 mm; greatest width submedian; L/W 1.7; vault a truncate paraboloid, with slightly curved sides and flattened summit; pelvis pyramidal with nearly straight to gently concave edges; V/P, 1.7; pelvic angle 75°–90°; ambulacra flat, with low, narrow rims; 8–9 transverse grooves in 3 mm; interambulacra flat to slightly concave; deltoids short ending 1–2 mm below summit.

Lower Chester, Rosiclare, Ill.; Monroe, St. Clair, Union, and Hardin cos., Ill.

This species is much like *P. symmetricus* (Ulrich, 1917, p. 245) with which it occurs but is slimmer with a smaller pelvic angle. It is much slimmer than *P. pyriformis*, *P. patei*, *P. abruptus*, and *P. altus*.

Plesiotype: Ind. Univ. Paleo. Coll., No. 5242, Paint Creek formation, Prairie du Long Creek, near Floraville, Ill.

## Pentremites kirki Hambach

(Pl. 5, figs. 10–12)

*Pentremites pyriformis* OWEN (not Say), 1842, Amer. Jour. Sci., vol. 43, p. 20, Fig. 3 (Chester, locality not stated)
*Pentremites kirki* HAMBACH, 1903, Trans. St. Louis Acad. Sci., vol. 13, p. 55, Pl. 5, fig. 18 (Lower Burlington limestone ??)
*Pentremites lyoni gracilens* ULRICH, 1917, Form. Chester Ser. West. Ky., Ky. Geol. Surv., p. 263, Pl. 7, figs. 30–32 (Golconda formation Golconda, Ill.)
Not *Pentremites lyoni gracilens* HAAS, 1945, Amer. Mus. Novitates, no. 1289, p. 4, Figs. 27, 36, 37

Calyx fusoidal, elongate bipyramidal, medium size, length 19–26 mm; width 9–13 mm; greatest width submedian; L/W, 2–2.1; vault long, narrow with strongly converging, nearly straight sides and rounded summit; pelvis pyramidal, elongate, with nearly straight to slightly convex sides;

V/P, 1.4–1.6; pelvic angle 50°–60°; ambulacra nearly flat, with low, narrow rims, 8–9 transverse grooves in 3 mm; interambulacra nearly flat to slightly concave; deltoids narrow, up to 5 mm long ending slightly below summit.

Golconda formation, Golconda, Ill.

Owen's figure is the first figure of a specimen assigned to *P. pyriformis*, but it is not typical, for Owen says (1842, p. 20, footnote), "The Pentremites here represented differs a little from the drawing of the *P. pyriformis* which I have seen."

This species is well characterized. It is probably derived from *P. welleri*. (*Cf.* Ulrich, 1917, Pl. 6, figs. 20, 21, 25, with Pl. 7, figs. 30, 32.) Its relationship to Ulrich's *P. lyoni* (1917, p. 262, Pl. 7, figs. 27–29), a synonym of *P. welleri*, is admitted. It is confined to the middle Chester, probably to the Golconda formation. Hambach's statement that his specimen came from the Burlington limestone seems impossible, but since he gave no locality the occurrence cannot be checked. There can be little doubt of the identity of *P. kirki* and *P. lyoni gracilens*.

Plesiotype: Ind. Univ. Paleo. Coll., No. 5243, Golconda formation, Lusk Creek, Pope Co., Ill.

## Pentremites turbinatus Hambach

(Pl. 5, fig. 13)

*Pentremites turbinatus* HAMBACH, 1903, Trans. St. Louis Acad. Sci., vol. 13, p. 54, Pl. 5, fig. 6 (Middle? Chester, Evansville, Ill.)

Calyx turbinate, small, 11 mm long and 9 mm wide; greatest width near top; L/W, 1.2; vault broadly convex, 4 mm high; pelvis funnel-shaped, 8 mm long, with sides slightly convex in profile; V/P, 0.5; pelvic angle 50°; ambulacra flat, without raised rims; transverse grooves about 11 in 3 mm; interambulacra "rather flat"; deltoids very short, "externally not visible" although 1.5 mm long in the figure, tips below summit.

Chester, probably middle, Evansville, Ill. "Very rare."

This form is obviously the young stage of some form with long pelvis, such as *P. pyriformis*, *P. patei*, *P. speciosus*, *P. girtyi* (*see* Ulrich, 1917, Pl. 7, figs. 11–13), or *P. okawensis*. The name is valid for a species, but until it is determined to which adult stage the young stage belongs, the species must remain unrecognizable.

## Pentremites speciosus Rowley

(Pl. 5, fig. 14)

*Pentremites speciosus* ROWLEY, *in* GREENE, 1903, Contrib. to Ind. Paleo., pt. 12, p. 119, Pl. 36, fig. 13 (Chester Group, Crittenden Co., Ky.)
*Pentremites elegans* WOOD (not Lyon), 1909, U. S. Nat. Mus. Bull. 64, p. 14, Pl. 2, figs. 10–12 (Sequatchie Valley and Sparta, Tenn.; Monte Sano, Ala.)

Calyx pyriform, doubly conical, medium size; measurements of the type, 25 mm long; 20 mm wide; greatest width supramedian; L/W, 1.3, vault conoidal, truncate, sides slightly convex, summit almost half as wide as the vault and slightly convex; length 11 mm; pelvis 14 mm long; V/P, 0.8; sides nearly straight except for a slight enlargement on the sides of the basals; pelvic angle 55°–60°; ambulacra almost flat, with low rims; 6–7 lateral grooves in 3 mm; interambulacra flat; deltoids 4 mm long, not quite even with spiracles.

Probably lower and middle Chester of Kentucky.

*Pentremites speciosus* is intermediate between *P. pyriformis* and *P. girtyi* and may be only a variety of either. The main distinction is in the smaller length/width ratio (1.3 instead of 1.5). This is expressed in the specimen by a slightly wider vault than that of *P. pyriformis* or *P. girtyi*.

## Pentremites girtyi Ulrich

(Pl. 1, fig. 8; Pl. 5, figs. 15, 16; Pl. 11, figs. 31–42)

*Pentremites girtyi* ULRICH, 1917, Form. Chester Ser. West. Ky., Ky. Geol. Surv., p. 261, Pl. 7, figs. 11–16 (Chester, Breckinridge, Todd, and Caldwell cos., Ky., Maxville limestone, Newtonville, Ohio, and Paint Creek formation in Monroe Co., Ill.)

Calyx pyriform, medium; length up to 26 mm, av. adult 20 mm; width 12–16 mm; greatest width supramedian; L/W, 1.5–1.6; vault paraboloid in adult specimens, truncate in young; pelvis an elongate pyramid with nearly straight sides; V/P, 0.7; pelvic angle 50°–60°; ambulacra flat or slightly concave, with low narrow margins; interambulacra nearly flat or gently concave; 7–8 lateral grooves in 3 mm; deltoids 2–3 mm long, almost reaching summit.

Lower and middle Chester, Kentucky, Ohio, and Illinois, Perry and Crawford cos., Indiana.

*Pentremites girtyi* was probably derived from *P. pyriformis*. (*See* discussion of *P. pyriformis*.) Perhaps *P. girtyi* represents a young adult and *P. okawensis* a more fully matured stage. This species eventually gave rise to *P. clavatus*, by further elongation of the pelvis.

Plesiotype: Ind. Univ. Paleo. Coll., No. 5244, Golconda limestone, half a mile north of Grantsburg, Ind.

### Pentremites okawensis Weller

(Pl. 5, figs. 17, 18; Pl. 11, figs. 43–54)

*Pentremites okawensis* WELLER, 1920, Ill. Geol. Surv., Bull. 41, p. 357, Pl. 10, figs. 5–7 (Golconda limestone, Randolph Co., Ill.; Golconda limestone, Pope and Hardin cos., Ill.; Glen Dean limestone, Hardin Co., Ill.).—BUTTS, 1926, Geol. Surv. Ala., Spec. Rept. 14, p. 198, Pl. 65, fig. 4 (Bangor limestone, Glen Dean horizon, Paint Rock River, Marshall Co., Ala.).—HAAS, 1945, Amer. Mus. Novitates, no. 1289, p. 7, Figs. 20–22 (Glen Dean limestone, Crane, Ind.)

Calyx pyriform, elongate, much higher than wide; length 25 mm or more; width 16 mm, position of greatest width supramedian at the ends of the ambulacra; L/W, 1.6; vault ovoid, with convex sides and summit; length of vault 10 mm; pelvis elongate, 15 mm long; V/P, 0.7–0.8, sides straight or very slightly concave; pelvic angle 50°–55°; ambulacra slightly convex with low rims; 7–8 lateral grooves in 3 mm; interambulacra flat; deltoids short, not well shown on type figure but up to 4 mm long on the writers' specimens, ending 1–2 mm below spiracles.

Golconda and Glen Dean limestone, Illinois, Alabama; Golconda limestone, near Shoals and Grantsburg, Indiana.

*Pentremites okawensis* is almost indistinguishable from *P. girtyi*. In general, *P. girtyi* has a slightly larger basal angle and is wider than *P. okawensis*. *P. clavatus* developed from *P. okawensis* by further elongation of the pelvis.

Plesiotypes: Ind. Univ. Paleo. Coll., No. 5245, Beaver Creek, half a mile southeast of Shoals, Ind.

### Pentremites clavatus Hambach

(Pl. 5, figs. 19–21)

*Pentremites clavatus* HAMBACH, 1880, Trans. St. Louis Acad. Sci., vol. 4, p. 157, Pl. B, fig. 5 (Chester limestone, Evansville, Ill.)

Calyx clavate, elongate, small to large; length from 12–32 mm, av. adult 22 mm; width 6–14 mm; av. adult 10 mm; greatest width supramedian; L/W, 2; vault ovoid with slightly flattened summit; pelvis long, triangular in lower part, pentagonal above, with nearly straight to slightly concave edges; V/P, 0.66, pelvic angle 35°; ambulacra slightly convex, with low narrow rims; 5–6 lateral grooves in 3 mm; deltoids very small, generally not visible especially on small specimens, always ending slightly below summit.

Kaskaskia limestone, Evansville, Ill., Golconda formation, Crawford and Martin cos., Ind.

This species represents the final developmental stage of *P. pyriformis*, with the pelvis elongated to its greatest degree, instead of the usual elongation of the vault. The small pelvic angle of the nepionic stage, common to all *Pentremites*, has been retained.

Plesiotypes: Ind. Univ. Paleo. Coll., Nos. 5246, 5246a, Golconda limestone, Beaver Creek, half a mile southeast of Shoals, Ind.

KEY TO SPECIES OF *PENTREMITES SULCATUS* GROUP

Page

1a. Basal angle 80°–90° in adult, 60°–80° in the young
  2a. Vault as long or longer than pelvis; V/P, 1–1.6
    3a. Sides of pelvis strongly concave—*P. springeri* Ulrich.............................. 62
    3b. Sides of pelvis slightly concave to straight
      4a. Length over width, 1—*P. angularis* Lyon.................................... 64
      4b. Length over width, 1.1–1.2
        5a. Deltoids short, not above summit
          6a. Radials without nodes—*P. elegans* Lyon................................. 64
          6b. Radials with nodes—*P. nodosus* Hambach................................ 65
        5b. Deltoids long, above summit—young *P. obesus* Lyon....................... 70
          syn. *P. spicatus altipelvis* Haas......................................... 70
  2b. Vault shorter than pelvis; V/P, 0.6–0.9—*P. calycinus* Lyon...................... 64
    equals young *P. elegans* Lyon.............................................. 64
1b. Basal angle 95°–120°
  2c. Deltoids even with mouth and spiracles
    3c. Interambulacra slightly concave
      4c. Basals without nodes
        5c. Vault semicircular in profile—*P. hambachi* Butts........................... 65
        5d. Vault a truncate pyramid—*P. cervinus* Hall.............................. 66
        5e. Vault paraboloid
          6c. Pelvic angle 95°–100°—*P. fohsi* Ulrich................................ 66
          6d. Pelvic angle 115°–120°—*P. robustus* Lyon.............................. 66
          syn. *P. fohsi marionensis* Ulrich....................................... 66
      4d. Basals with nodes
        5f. Vault with slightly convex sides
          6e. Deltoids do not flare—*P. tulipaformis* Hambach......................... 67
          6f. Deltoids flare outwardly—*P. laminatus* Easton........................... 67
        5g. Vault strongly convex—*P. gutschicki* n. sp................................ 68
    3d. Interambulacra strongly concave
      4e. Length and width subequal; L/W, 0.9–1.1
        5h. Radials without spines
          6g. Summit concave—*P. sulcatus* Roemer................................. 68
          6h. Summit convex—*P. hemisphericus* Hambach............................ 68
        5i. Radials with spines—*P. spinosus* Hambach.............................. 69
      4f. Length much greater than width; L. W, 1.3–1.4
        5j. Summit concave—*P. chesterensis* Hambach.............................. 69
        5k. Summit truncate—*P. maccalliei* Schuchert.............................. 70
  2d. Deltoids extend above mouth and spiracles
    3e. Basal plates not inflated
      4g. Basal angle 90°–95°—*P. obesus* Lyon..................................... 70
        syn. *P. spicatus altipelvis* Haas........................................ 70
      4h. Basal angle 100°–125°—*P. spicatus porrectus* Haas........................ 73
    3f. Basal plates inflated—*P. broadheadi* Hambach.............................. 71
1c. Basal angle 125°–165°
  2e. Deltoids extend above mouth and spiracles
    3g. Deltoids flare outward
      4i. Ambulacra broad V-shaped—*P. halli* n. sp................................ 71
      4j. Ambulacra narrow V-shaped—*P. cherokeeus* Troost........................ 72
    3h. Deltoids do not flare outward
      4k. Basal angle 125°–140°

Page

51. Ambulacral flange not serrate
    6i. Vault-pelvis ratio, 0.7–0.8—*P. gutschicki* n. sp. .......................... 68
    6j. Vault-pelvis ratio, 1—*P. spicatus* Ulrich. ........................ 72
    6k. Vault-pelvis ratio, 2—*P. spicatus porrectus* Haas. .................... 73
  5m. Ambulacral flange serrate—*P. serratus* Hambach. .................... 73
  4l. Base flat, angle 160° or more—*P. basilaris* Hambach. ................... 74
2f. Deltoids do not reach summit—*P. bradleyi* Meek. ...................... 74

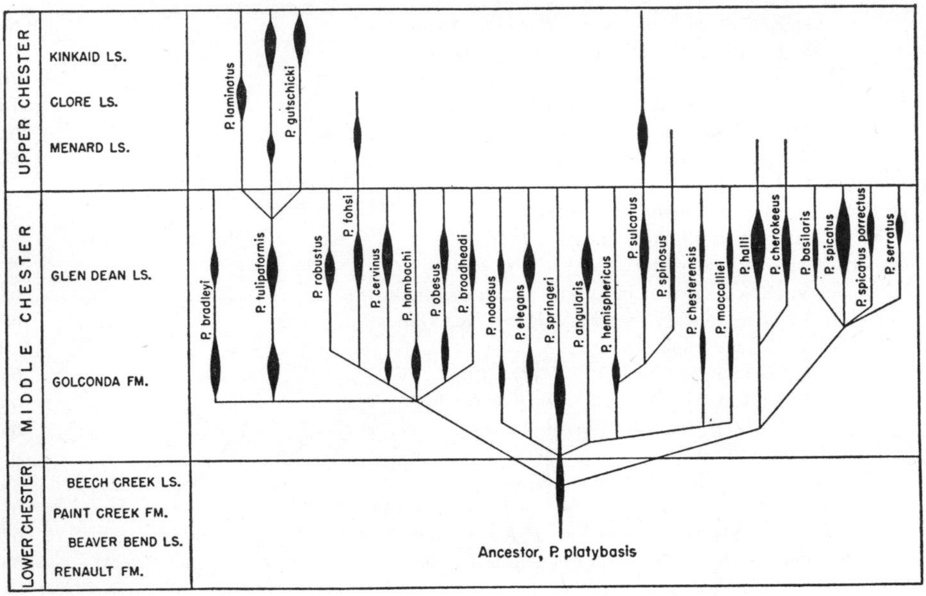

FIGURE 5.—*Geologic occurrence and phylogeny of the* Pentremites sulcatus *group*

## DESCRIPTION OF SPECIES OF *PENTREMITES SULCATUS* GROUP

### Pentremites springeri Ulrich

(Pl. 5, figs. 22–24)

*Pentremites springeri* ULRICH, 1917, Form. Chester Ser. West. Ky., Ky. Geol. Surv., p. 255, Pl. 5, figs. 36–40 (Paint Creek formation, New Bethel, Ky.)

Calyx subpentameral in side view, length and width subequal, up to 19 mm, greatest width submedian; L/W, 1–1.3; vault with moderately curved sides and flat summit, 1 mm longer than pelvis; pelvis with concave sides, with one columnal attached to stem in many specimens; V/P, 1.1–1.6, pelvic angle 80°–90°; ambulacra flat at point and concave at top, with low, narrow rims; 9 transverse grooves in 3 mm; interambulacra moderately concave; deltoids less than half the length of ambulacra, up to 5 mm in length, extending to level of mouth.

Lower Chester, rare in the Paint Creek formation of Kentucky; Middle Chester, common in the Golconda formation at Shoals, Sulphur, Crawford Co., Gerald, Perry Co., Ind., common in the Golconda formation at Golconda, Ill.

This species is the earliest species of the *P. sulcatus* group; consequently it is small and generalized. It is characterized by concave sides of the base and bulbous vault. It is closely related to *P. angularis* Lyon and *P. elegans* Lyon, differing mainly in the concave edges of the pelvis. *P. springeri* gave rise to *P. obsesus*, as Ulrich (1917, p. 255) suggested, but *P. hambachi* was the intermediate stage.

TABLE 5.—*Main characters of the* Pentremites sulcatus *group*

Calices ovoid, large; vault 0.6–7 times pelvis; pelvic angle 80°–165°; ambulacra wide, strongly concave, and flanged; interambulacra moderately to strongly concave; deltoids even with or above summit. One in the lower Chester, common in the middle Chester, few in the upper Chester. (Synonyms in italics)

| Species | Length Width | Vault Pelvis | Pelvic angle | Ambulacra | Inter-ambulacra | Original horizon and locality |
|---|---|---|---|---|---|---|
| P. springeri | 1–1.3 | 1.1–1.6 | 80°–90° | concave | concave | Paint Creek; Ky. |
| P. angularis | 1 | 1 | 85°–95° | concave | concave | Glen Dean; Ky. |
| P. elegans | 1.2 | 1.1–1.2 | 80°–90° | slightly concave | slightly concave | Glen Dean; Ky. |
| *P. calycinus* | 1 | 0.6 | 80° | concave | slightly concave | Glen Dean; Ky. |
| *P. canalis* | 1.3 | 1.6 | 85° | concave | nearly flat | Glen Dean; Ky. |
| *P. praematurus* | 1.3 | 0.6 | 75° | concave | flat | Ste. Genevieve; Ill. |
| P. nodosus | 1.2 | 0.8 | 80°–85° | slightly concave | slightly concave | Middle Chester; Ill. |
| P. hambachi | 1 | 2–3 | 95°–100° | concave | slightly concave | Glen Dean; Ala. |
| P. cervinus | 1 | 1.7 | 100°–110° | moderately concave | slightly concave | Middle Chester; Ill. |
| P. fohsi | 1.1–1.2 | 1.7 | 95°–100° | slightly concave | flat | Glen Dean; Ky. |
| P. robustus | 1.1 | 2.5–4 av. 2.8 | 115°–120° | concave | slightly concave | Glen Dean; Ky. |
| *P. fohsi marionensis* | 1.1 | 4 | 120° | concave | slightly concave | Glen Dean; Ky. |
| P. tulipaformis | 1 | 2–4 | 100°–135° | slightly concave | slightly concave | Middle Chester; Ill. |
| P. laminatus | 1.1 | 3.3 | 115° | broad V-shaped | slightly concave | Late Chester; Ark. |
| P. gutschicki | 0.7–0.8 | 5.6–8 | 135° | deeply concave | moderately concave | Kinkaid; Ky. |
| P. hemisphericus | 1 | 2.5 | 120° | strongly concave | strongly concave | Golconda; Ill. |
| P. sulcatus | 1 | 4 | 115°–125° | strongly concave | strongly concave | Glen Dean; Ill. |
| P. spinosus | 0.9 | 2.6 | 115° | strongly concave | strongly concave | Glen Dean; Ill. |
| P. chesterensis | 1.3 | 3 | 110° | strongly concave | strongly concave | Middle Chester; Ill. |
| P. maccalliei | 1.4 | 2.8–3 | 100°–115° | concave | strongly concave | Middle Chester; Ga. |
| P. obesus | 1.2 | 1.5 | 90°–95° | strongly concave | concave | Golconda; Ky. |
| *P. spicatus altipelvis* | 1.2 | 2 | 90° | strongly concave | concave | Glen Dean; Ind. |
| P. broadheadi | 1.0 | 2 | 100°–110° | strongly concave | strongly concave | Chester; Ill. |

TABLE 5.—*Continued*

| Species | Length Width | Vault Pelvis | Pelvic angle | Ambulacra | Inter-ambulacra | Original horizon and locality |
|---------|------|------|--------------|-----------|----------------|------------------------------|
| P. halli | 0.9–1 | 1.4–2.5 | 100°–130° | concave | strongly concave | Glen Dean; Ky. |
| P. cherokeeus | 1 | 3 | 140° | deep, narrow | V-shaped | Middle Chester; Tenn. |
| P. spicatus | 1–1.1 | 5 | 120°–140° | deep | strongly concave | Glen Dean; Ky. |
| P. spicatus porrectus | 1.2 | 3 | 100°–125° | strongly concave | strongly concave | Glen Dean; Ind. |
| P. serratus | 1 | 4 | 125° | deep | strongly concave | Glen Dean; Ill. |
| P. basilaris | 1.1 | 7 | 165° | strongly concave | strongly concave | Glen Dean; Ill. |
| P. bradleyi | 1 | 3–5 | 135°–145° | concave | slightly concave | Upper Mississippian; Mont. |

Plesiotype: Ind. Univ. Paleo. Coll.; No. 5247, Golconda formation, half a mile north of Grantsburg, Ind.; No. 5248, Golconda formation, Lusk Creek, Golconda, Ill.

### Pentremites angularis Lyon

(Pl. 5, figs. 25–27)

*Pentremites angularis* LYON, 1860, Trans. St. Louis Acad. Sci., vol. 1, no. 4, p. 631, Pl. 20, fig. 3a–c (Glen Dean limestone, Falls of Rough Creek, Breckinridge Co., Ky.)

Calyx broad pyriform; length and width nearly equal, 16–22 mm; greatest width submedian in the adult stage; L/W, 1; vault with slightly convex sides and broad, flat summit; length of vault 6–11 mm; pelvis 10–11 mm long; V/P, 0.7–1; sides nearly straight; pelvic angle 85°–95°; ambulacra strongly concave, with low rims; 9 lateral grooves in 3 mm; interambulacra slightly concave; deltoids up to 5 mm long, even with summit and slightly flaring.

Middle Chester, Kentucky; abundant in the Glen Dean limestone near Cloverport, Ky., and Rome, Ind.

The basal angle follows the usual evolutionary trend in this group, becoming larger as the individual becomes older, and the vault lengthens with age. *P. angularis* is distinguished from *P. elegans* by its broader vault, larger basal angle, more concave ambulacra, and flaring deltoids, and from *P. springeri* by the straight edges of the pelvis.

Plesiotype: Ind. Univ. Paleo. Coll.; Nos. 5249, 5249a, Glen Dean limestone, 1 mile east of Cloverport, Ky.

### Pentremites elegans Lyon

(Pl. 5, figs. 28–30; Pl. 6, figs. 1–4)

*Pentremites elegans* LYON, 1860, Trans. St. Louis Acad. Sci., vol. 1, no. 4, p. 632, Pl. 20, fig. 4a–c (Glen Dean limestone, near Grayson Springs and north of Litchfield, Grayson Co., Ky.).— WHITFIELD, 1893, Rept. Geol. Surv. Ohio. vol, 7, p. 466, Pl. 9, fig. 4 (Maxville limestone, Newtonville, Ohio)
*Pentremites calycinus* LYON, 1860, Trans. St. Louis Acad. Sci., vol. 1, no. 4, p. 628, Pl. 20, fig. 1a–c (Golconda limestone, near Eskridge's Ferry, Grayson Co., Ky.).—ROWLEY, *in* GREENE, 1903, Contrib. to Ind. Paleo., pt. 12, p. 125, Pl. 36, figs. 39–40 (Chester limestone, Clifty Sta., Hardin Co., Ky.)
*Pentremites canalis* ULRICH, 1917, Form. Chester Ser. West. Ky., Ky. Geol. Surv., p. 262, Pl. 7, figs. 23, 24, 25, 26 (Glen Dean limestone, Sloans Valley, Ky.)

*Pentremites praematurus* ULRICH, 1917, Form. Chester Ser. West. Ky., Ky. Geol. Surv., p. 245, Pl. 2, fig. 20 (Ste. Genevieve limestone, Ill.)

Calyx pyriform, short to medium, up to 25 mm long, av. adult 20 mm; width up to 22 mm, av. adult 17 mm; greatest width submedian; L/W 1.2–1.3; vault with broad, paraboloid profile, up to 17 mm long, av. adult 13 mm; pelvis up to 10 mm long, av. adult 8 mm, with nearly straight sides; V/P, 1.1–1.2; pelvic angle 80°–90°; ambulacra moderately concave, 4–5 mm wide, with low narrow rims; 7–9 transverse grooves in 3 mm, most at broad end of ambulacrum; interambulacra nearly flat to slightly concave; deltoids short, extending almost to summit and not flaring.

Middle Chester: Golconda not common, abundant in the Glen Dean limestone, Grayson and Hardin cos., Ky., Sloans Valley, Ky., Cloverport, Ky., and Rome, Ind.; Maxville limestone, Ohio.

*Pentremites elegans* is one of the early described species which have been largely ignored by later workers. This is difficult to explain since Lyon's figure is one of the better illustrations, and the species is abundant in the Glen Dean limestone. This species is similar to some members of the *P. pyriformis* group, such as *P. welleri*, *P. abruptus*, and *P. pyramidatus*, differing markedly in having concave ambulacra. *P. canalis* is identical with *P. elegans* in all respects, including geologic range. *P. praematurus* is obviously a young specimen and may belong in the *P. pyriformis* group, but Ulrich states (1917, p. 245), "the ambulacral areas are more deeply excavated than in *P. girtyi*", and the figure shows concave ambulacra, which should ally the form to the *P. sulcatus* group. *P. calycinus* is a young specimen with a columnal still attached.

Plesiotype: Ind. Univ. Paleo. Coll.; No. 5250, Glen Dean formation, 1 mile east of Cloverport, Ky.

## Pentremites nodosus Hambach

### (Pl. 6, figs. 5, 6)

*Pentremites nodosus* HAMBACH, 1880, Trans. St. Louis Acad. Sci., vol. 4, p. 155, Pl. B, fig. 2. (Chester limestone, Randolph Co., Ill.).—WELLER, 1920, Ill. Geol. Surv., Bull. 41, p. 356, Pl. 4, fig. 25, (Lower Okaw limestone or Golconda limestone, Randolph Co., Ill.)

Calyx subovoid, large; length 23–26 mm; width 20–23 mm; greatest width median; L/W, 1.1–1.2, vault subhemispherical, with narrow, rounded summit; length of vault 11–14 mm; pelvis averages 13 mm long; V/P, 1–1.2, sides of pelvis slightly concave; pelvic angle 80°–85°; ambulacra slightly depressed, narrow in the type; ambulacral rims low and narrow; 8–10 transverse ridges in 3 mm; interambulacra slightly concave; node on each tip of radials, just below deltoid suture; deltoids 5–7 mm long, even with spiracles.

Rare in the Golconda formation in Illinois; not found in Indiana or Kentucky.

The radial nodes and smaller pelvic angle distinguish this species from *P. hambachi*. Weller (1920, p. 357) states, "The roughness of callous development upon the interambulacral regions of this species . . . is a character that is somewhat inconstant, being nearly obsolete in some individuals. It is quite likely that the feature increases in prominence with the age of the individuals, and that younger specimens may be quite free from it." An analogous condition occurs in the basal callosities of *P. tulipaformis* and *P. gutschicki*. The spines on lower parts of the radials in *P. spinosus* Hambach is a similar condition, pathologic, teratologic, or gerontic.

## Pentremites hambachi Butts

### (Pl. 6, figs. 7, 8)

*Pentremites hambachi* BUTTS, 1926, Geol. Surv. Ala., Spec. Rept. No. 14, p. 198, Pl. 65, fig. 2. (Bangor limestone, Glen Dean horizon, Paint Rock River, Marshall Co., Ala.)

Calyx ovoid, length and width subequal, length up to 31 mm, width up to 26 mm, greatest width submedian, L/W, 1.0–1.4 mm, approaching maximum as size of specimen increases; vault and summit rounded, nearly hemispherical, length up to 21 mm; pelvic profile triangular, low, broad, length up to 10 mm, sides very slightly concave; V/P, 2–3; pelvic angle 95°–100°; ambulacra concave, the two sides convex, with low narrow rims; 7–9 transverse grooves in 3 mm; interambulacra slightly concave; deltoids up to 9 mm long, ending near summit.

Common in the middle Chester of Alabama and Indiana. Expectable in all other States where pentremites have been found in the middle part of the Chester Series.

The first recording of *Pentremites hambachi* was from the Glen Dean. Specimens from the Golconda type section, like Butts' specimen, lack the deeply concave ambulacra and the projecting spikelike deltoids which characterize many Glen Dean species.

Pleisotype: Ind. Univ. Paleo. Coll.; No. 5251, Golconda formation, Lusk Creek, North of Golconda, Ill.

## Pentremites cervinus Hall

### (Pl. 6, figs. 9, 10)

*Pentremites cervinus* HALL, 1858, Rept. Geol. Surv. Iowa, vol. 1, pt. 2, p. 690–691, Pl. 25, fig. 11 a–b (Chester limestone, Chester, Ill., Huntsville, Ala.)

Calyx pentameral in side view, large; length and width almost equal, from 18–25 mm in either direction; greatest width submedian; L/W, 1; vault with nearly straight sides, up to 15 mm long; pelvis 7–10 mm long, with straight sides; V/P, 1.7; pelvic angle 100°–110°; ambulacra averaging 4 mm wide, moderately concave, with low margins; 8–9 lateral grooves in 3 mm; interambulacra slightly concave; deltoids small and short, about 5 mm, ending near summit.

Middle Chester, Chester, Ill., Huntsville, Ala.; Golconda formation, Grantsburg, Ind.; Glen Dean limestone, Cloverport, Ky., Rome, Ind.

Hall noted the resemblance to *P. sulcatus* Roemer and pointed out the more prominent base (smaller angle) and rapidly converging vault. It is similar to *P. hambachi*, but the vault of *P. cervinus* has straighter sides and *P. hambachi* has a larger V/P ratio. Specimens studied do not have "a little protuberant just without the column area" (Hall, 1858, p. 690).

Pleistotype: Ind. Univ. Paleo. Coll.; No. 5252, Golconda formation, half a mile north of Grantsburg, Ind.

## Pentremites fohsi Ulrich

### (Pl. 6, figs. 11–13; Pl. 13, fig. 6)

*Pentremites fohsi* ULRICH, 1905, U. S. Geol. Surv., Prof. Pap. 36, p. 64, Pl. 7, figs. 5–9 (Middle and Upper Chester, Princeton, Marion, Ky.).—BUTTS, 1917, Miss. Form. of West. Ky., Ky. Geol. Surv., p. 101, Pl. 24, fig. 21 (Glen Dean limestone, Marion, Ky.).—WELLER, 1920, Ill. Geol. Surv., Bull. 41, p. 370–371, Pl. 10, fig. 4 (Menard limestone, Union, Johnson, Pope cos., Ill.)

Calyx ovoid, very large, length up to 41 mm, width up to 35 mm; greatest width submedian, L/W, 1.1–1.2; vault paraboloid with flattened summit, length up to 25 mm; pelvis up to 16 mm long; V/P, 1.6; sides of pelvis almost straight, with basals slightly bulging; pelvic angle 95°–100°; ambulacra concave, though shallow, with low, narrow rims; 8 transverse grooves in 3 mm; interambulacra flat with slight concavity along suture; deltoids up to 12 mm long, not extending to summit.

Rare to common in the middle and upper Chester of Kentucky, Illinois, and Indiana.

*Pentremites fohsi* is a large form closely related to *P. cervinus*; the most apparent difference is the increase in size and V/P ratio. *P. robustus* Lyon is a direct descendant of *P. fohsi*, and the variety *P. fohsi marionensis* is identical with *P. robustus*.

Plesiotype: Ind. Univ. Paleo. Coll.; No. 5253, Glen Dean limestone, Sloans Valley, Ky.

## Pentremites robustus Lyon

### (Pl. 6, figs. 14, 15)

*Pentremites robustus* LYON, 1860, Trans. St. Louis Acad. Sci., vol. 1, no. 4, p. 629, Pl. 20, fig. 2 a–c (Glen Dean limestone, Grayson Co., Ky.)

Not *Pentremites robustus* ROWLEY, *in* GREENE, 1903, Contrib. to Ind. Paleo., p. 118, Pl. 36, fig. 12 = *P. biconvexus* Ulrich

*Pentremites fohsi marionensis* ULRICH, 1905, U. S. Geol. Surv., Prof. Paper 36, p. 64, Pl. 7, figs. 10–11 (Middle and Upper Chester, Marion, Ky.)

Calyx ovoid or globular, large, length up to 36 mm; width up to 33 mm; greatest width sub-median; L/W, 1.1; vault subspheroidal, sides strongly curved, summit narrow and slightly convex; pelvis short, up to 8 mm long; V/P, 3–4; sides of pelvis concave to sigmoid; pelvic angle 115°–120°; ambulacra concave, not deeply excavated; transverse grooves about 11 in 3 mm; rims low, without prominent flanges; interambulacra moderately concave; deltoids long, up to 13 mm, even with or above summit.

Rare in the middle Chester of Kentucky.

This species may be the ephebic stage of *P. fohsi* Ulrich. The basal angle increases progressively from *P. fohsi* to *P. robustus*, as does the vault to pelvis ratio.

The specimen figured by Rowley (1903, p. 118, Pl. 36, fig. 12) does not belong to *P. robustus* since it is not wide enough and does not have the depressed ambulacra; it closely resembles *P. biconvexus*.

## Pentremites tulipaformis Hambach

### (Pl. 6, figs. 16, 17)

*Pentremites tulipaformis* HAMBACH, 1903, Trans. St. Louis Acad. Sci., vol. 13, Pl. 4, figs. 10, 11 (Chester limestone, Kaskaskia, Ill.).—BUTTS, 1917, Miss. Forms. West. Ky., Ky. Geol. Surv., p. 100, Pl. 24, fig. 5 (Glen Dean limestone, Caney Creek, 5 miles southeast of Cloverport, Ky.)
*Pentremites brevis* WELLER (not Ulrich), 1920, Ill. Geol. Surv., Bull. 41, p. 369, Pl. 4, figs. 43–46 (Glen Dean limestone, Randolph and Hardin cos., Ill.).—HAAS (not Ulrich, but Weller), 1945, Amer. Mus. Novitates, no. 1289, p. 8, Figs. 32, 34, 39, 40 (Glen Dean limestone, Crane, Ind.)

Calyx ovoid, subpentameral in side view, medium size, adult 17–25 mm long and 18–23 mm wide, greatest width suprabasal; L/W, 1–1.1; vault with nearly straight sides and truncate to slightly concave summit, 11–20 mm high; pelvis short, 3–6 mm high, with concave sides down to the nodes; V/P, 2–4; basal plates prominently nodose; stem facet in depression; pelvic angle 100°–135°, averaging about 120°, not including basal nodes; ambulacra deeply concave, with narrow shallow median groove, and low, narrow ambulacral rims; 9 transverse grooves in 3 mm; interambulacra slightly concave; deltoids 5–10 mm long, 5–8 mm wide, extending about half a mm above the mouth, but not flaring outward.

Abundant in the Golconda, 1 mile south of Gerald, Perry Co., Ind. Rare in the Glen Dean limestone of Kaskaskia, Ill. Abundant in the Glen Dean limestone, 1 mile east of Cloverport, Ky., and 1 mile west of Rome, Ind., and in Randolph and Hardin cos., Ill. Abundant in the late Chester, Kinkaid formation, 8 miles northeast of Marion, Ky., the P.W.A. quarry, Cave Hill, Equality Quad., Ill., and 2 miles west of Robbs, Ill. Kinkaid specimens collected by Dr. R. C. Gutschick, University of Notre Dame.

This species is one of the most abundant forms. It occurs in Kentucky and Illinois with *P. gutschicki*. Except for the strongly convex vault of *P. gutschicki* the two species are very similar, including the basal nodes.

Plesiotype: Ind. Univ. Paleo. Coll.; No. 5254, Alexander Stone Co. quarry, 8 miles northwest of Marion, Ky.

## Pentremites laminatus Easton

### (Pl. 7, fig. 1)

*Pentremites laminatus* EASTON, 1943, Jour. Paleo., vol. 17, p. 136, Pl. 21, figs. 9, 10 (Pitkin formation, 0.3 miles south of Leslie, Ark.)

Calyx ovoid, medium size, about 15 mm long and 13 mm wide (measured from type figure); greatest width suprabasal; L/W, 1.1–1.2; sides of vault moderately convex, summit truncate, length 11.5 mm; pelvis about 3.5 mm high, with straight profile down to basal plates where it is convex; basal plates each with node; stem facet a little higher than adjacent nodes; V/P, 3.3; pelvic angle 115°, measured on radials; ambulacra wide, about 4 mm at top, broad V-shaped in cross section; ambulacral rims low and narrow; lateral grooves 9–10 in 3 mm; interambulacra moderately concave; deltoids about 5 mm long, flaring outward moderately at tips, even with or above mouth.

Pitkin formation, late Chester, Leslie, Ark.

This species seems to be identical with *P. tulipaformis* in all respects except for the flaring deltoids. The "*P. brevis*" in the Walker Museum, apparently the specimens described and figured by Weller (1920, p. 369, Pl. 4, figs. 43–46), with which *P. laminatus* is compared, is typical *P. tulipaformis*. (*P. brevis* Ulrich is a synonym of *P. godoni abbreviatus* Hambach.) The growth laminations which obviously suggested the name for this species are common to all well-preserved pentremites, as is the triangular area on the interambulacrum described by Easton (1943, p. 137).

## Pentremites gutschicki Galloway and Kaska, n.sp.

### (Pl. 7, figs. 2, 3; Pl. 12, figs. 13–21)

Calyx globular, rather large, 18–20 mm long and 23–25 mm in diameter; greatest width suprabasal; L/W, 0.7–0.8; vault strongly convex, 16–17 mm high, with concave summit; pelvis short, 2–5 mm, broad conical, with concave edges and nodose basal plates; V/P, 5.6–8; pelvic angle about 135°, 90°–120° in young specimens; ambulacra about 4 mm wide at midlength, rather deeply and evenly concave, with narrow and shallow median groove, 9–10 transverse grooves in 3 mm, closer at broad end; ambulacral rims absent to slightly elevated and narrow; interambulacra strongly concave at lower end to nearly flat at upper end; deltoids large, about 10 mm long (measured parallel to surface) or half the height of vault, and about 8 mm wide, upper tips extending about 1 mm above mouth, but not flaring radially outward.

Latest Chester, Kinkaid formation, Alexander Stone Co. quarry, 8 miles northeast of Marion, Ky. Abundant.

This species resembles *P. sulcatus* but differs in the U-shaped instead of V-shaped ambulacra, in the less concave interambulacra, the more convex vault, and in the nodose basal plates. *P. tulipaformis*, occurring with *P. gutschicki*, has nodose basals, but the sides of the vault are nearly straight instead of highly convex.

Holotype: Pl. 7, fig. 2; Pl. 12, fig. 21. Ind. Paleo. Coll. No. 5255; paratypes: Pl. 7, fig. 3; Pl. 12, figs. 13–20. No. 5255a, Kinkaid formation, Alexander Stone Co. quarry, 8 miles northwest of Marion, Ky.

Collected by Dr. R. C. Gutschick, University of Notre Dame.

## Pentremites hemisphericus Hambach

### (Pl. 7, figs. 4, 5)

*Pentremites hemisphericus* HAMBACH, 1880, Trans. St. Louis Acad. Sci., vol. 4, p. 157, Pl. B, fig. 7 (Chester limestone, Chester, Ill., common; Evansville, Ill., rare)

Calyx subhemispherical, very large; length 36 mm; width 33 mm; position of greatest width suprabasal; L/W, 1.1; vault hemispherical with very convex sides and rounded summit; length 26 mm; pelvis 10 mm long, with sigmoid sides and prominent basal plates; V/P, 2.6, pelvic angle 120°; ambulacra very broad, slightly depressed, each half convex; about 10 transverse grooves in 3 mm; ambulacral rims moderately high; interambulacra strongly concave; deltoids long, up to 13 mm, even with spiracles.

Rare to common in the Golconda formation of Illinois.

*Pentremites hemisphericus* is a very robust and globular species. It is similar to *P. sulcatus* and may be identical, lacking only the slightly flaring deltoids and concave summit. Hambach (1880, p. 158) says *P. hemisphericus* differs from *P. robustus* only in having fewer transverse grooves in the same space.

Plesiotype: Ind. Univ. Paleo. Coll.; No. 5256, Golconda formation, north of Golconda, Ill.

## Pentremites sulcatus (Roemer)

### (Pl. 7, figs. 6, 7; Pl. 12, figs. 1–12)

*Pentatrematites sulcatus* ROEMER, 1851, Archiv. fur Naturgeschichte, Jahrg. 17, Bd. 1, p. 323, Pl. 6, figs. 10 a–c (Prairie du Long, south of Belleville, Ill.).—ROEMER, *in* BROWN, 1853, Lethaea Geognostica, Stuttgart, Theil 2, p. 282, Pl. 4, fig. 9 a–b (Monte Sano, Huntsville, Ala.).—

ROEMER, 1876, Lethaea Paleozoica, Stuttgart, Pl. 41, Fig. 1 a-c (Prairie du Long, near Belleville, Illinois. Reproduction of original figure)
*Pentremites sulcatus* SHUMARD, 1858, Trans. St. Louis Acad. Sci., vol. 1, no. 2, p. 243, 246 (Chester, Long Prairie, Ill.; Sparta, Crab Orchard, Tenn.; Mt. Sano, Ala.; Perry Co., Mo.).—ETHERIDGE AND CARPENTER 1886, Cat. Blastoidea Geol. Dept. British Mus. (Nat. Hist.), London, p. 165, Pl. 1, figs. 8–10; Pl. 2, fig. 31; Pl. 16, fig. 20; Pl. 18, fig. 5 (Chester, Ill.).—HAMBACH, 1903, Trans. St. Louis Acad. Sci., vol. 13, p. 39, Pl. 6, figs. 1–12 (Locality not given)

Calyx subglobose, very large; length 35 mm, width 36 mm; position of greatest width median; L/W, 1; vault subhemispherical, with strongly convex sides and slightly concave, narrow summit, length 28 mm; V/P, 4; pelvis short, length 7 mm, with sigmoidal sides and raised basal plates; pelvic angle 115°–125°; ambulacra concave, forming a broad flat V, with moderately high rims; 8 transverse grooves in 3 mm; interambulacra strongly concave, making basal view appear almost star-shaped; deltoids 10 mm long extending about 2 mm above spiracles, neither flaring nor high and pointed.

Common in the Glen Dean limestone, and upper Chester of Illinois, Indiana, Kentucky, Tennessee, Alabama, and Missouri.

*Pentremites sulcatus* is a form which has often been misidentified. *P. broadheadi* is very similar to it, except that it possesses swollen basals and extended deltoids. *P. spicatus* which has long deltoids has also been confused with *P. sulcatus*. Hambach (1903, Pl. 6) figured what he considered an ontogenetic series of *P. sulcatus* but it may be a series of *P. spicatus* (Pl. 12, figs. 1–12). Haas (1945, p. 1) stated that *P. spicatus* has usually, though wrongly, been referred to *P. sulcatus*, which lacks the protruding spikelike deltoids. The distinction between the two species is well shown by the type figures of each. *P. sulcatus* has V-shaped rather than U-shaped ambulacra, only moderately, rather than very high ambulacral rims, only slightly protruding, rather than spikelike deltoids, and a sigmoid, rather than doubly concave pelvis. Examination of specimens at the U. S. National Museum confirmed these determinations.

*P. sulcatus* seems to be most abundant in the Menard limestone, where many of the specimens are thinner and vary otherwise from the type figure. A specimen from Prof. Courtney Werner's collection, which is almost exactly like Roemer's type figure, is shown on Pl. 7, fig. 7.

### Pentremites spinosus Hambach

#### (Pl. 7, fig. 8)

*Pentremites spinosus* HAMBACH, 1880, Trans. St. Louis Acad. Sci., vol. 4, p. 154, Pl. B., fig. 1 (Chester limestone, Chester, Ill.)

Calyx medium, subglobose, length 18 mm, width 20 mm, greatest width median, L/W, 0.9; vault with stronly convex sides and broad flat summit; length 13 mm, pelvis 5 mm long, with small projecting spines up to 3 mm long on each side of the radial plates near base of adjoining ambulacra; V/P, 2.6; sides of pelvis nearly straight to slightly concave; pelvic angle 115°; ambulacra broad, concave, with high rims; 7 lateral grooves in 3 mm; interambulacra concave; deltoids 5 mm long, projecting above spiracles.

Very rare, reported only once from the Chester of Illinois.

This species is distinguished from *Pentremites sulcatus* by the small projecting spines at the bases of the radial plates. Such a form may be a teratologic or pathologic specimen, or the spines may represent a gerontic or phylogerontic condition, as do the interambulacral nodes of *P. nodosus*.

### Pentremites chesterensis Hambach

#### (Pl. 8, fig. 1)

*Pentremites chesterensis* HAMBACH, 1880, Trans. St. Louis Acad. Sci., vol. 4, p. 158, Pl. B, fig. 9 (Chester limestone, Randolph Co. and Chester, Ill.)
Not *Pentremites chesterensis* ROWLEY, *in* GREENE, 1903, Contrib. to Ind. Paleo., vol. 1, pt. 12, p. 117, Pl. 36, figs. 7–8, not 9–11 (Chester limestone, Big Clifty, Hardin Co., Ky., and 5 miles northwest of Bowling Green, Ky.)

Calyx ovoid, large; length 28 mm; width 22 mm; position of greatest width submedian; L/W, 1.3; vault paraboloid, with sides becoming more convex toward summit; summit narrow, slightly concave; length 21 mm; pelvis 7 mm long; V/P, 3; sides of pelvis sigmoid; basal plates protuberant; pelvic angle 110°; ambulacra narrow and depressed, with moderate rims; interambulacra deeply concave; 7 lateral grooves in 3 mm; deltoids 7–9 mm long, even with spiracles.

Reported only from the Chester of Illinois and Kentucky, probably middle Chester.

*Pentremites chesterensis* is a rare species. Rowley (1903, p. 117) has the only reference to it after the original description, but this form seems to have convex ambulacra. *P. chesterensis* is similar to *P. spicatus porrectus* Haas in general outline but lacks the high ambulacral rims and protruding spikelike deltoids which characterize those forms developed out of *P. halli*. *P. maccalliei* may be a synonym of *P. chesterensis*; the writers were unable to obtain a specimen of *P. chesterensis* for comparison.

## Pentremites maccalliei Schuchert

### (Pl. 8, figs. 2, 3)

*Pentremites maccalliei* SCHUCHERT, 1906, Proc., U. S. Nat. Mus., vol. 30, No. 1467, p. 759–760, Figs. 1–3 (Bangor limestone, Cole City, Ga.)
*Pentremites maccalliei* BUTTS, 1941, Va. Geol. Surv., Bull. 52, p. 250, Pl. 132, figs. 26–27 (Bluefield shale, Golconda; Glenlyn Mercer Co., W. Va.) (Misspelled "maccalleyi" on Plate)
*Pentremites gianteus* ALLEN AND LESTER, 1953, Ga. Geol. Surv. Bull. 60, p. 192, Pl. 4, fig. 1; Pl. 5, Pl. 6, figs. 5, 6 (not figs. 1–4) (Gasper-Ste. Genevieve zone, Little Sand Mountain, Ga); 1954, Bull. 62, p. 113, Pl. 29, figs. 27–29 (same locality and horizon)

Calyx conoidal, very large; 50–54 mm long, width 37–40; greatest width suprabasal; L/W, 1.3–1.4; vault 40–45 mm long, with moderately curved sides and nearly flat summit; pelvis 12–15 mm long, with sigmoid profile; basal plates protuberant; V/P, 2.8–3; pelvic angle, 100°–115°; ambulacra broad, about 9 mm, broadly V-shaped at upper ends, with slightly elevated rims, each side moderately convex; 7–8 transverse grooves in 3 mm, closer at wide end, 110 in an ambulacrum 39 mm long; interambulacra deeply concave, broadly V-shaped; deltoids 15–22 mm long, reaching summit.

Bangor limestone, Cole City, Ga.; Gasper-Ste. Genevieve argillaceous limestone, Little Sand Mountain, Ga.; Bluefield shale (Golconda), Mercer Co., Va.

*Pentremites maccaliei* is an unusually large species, restricted to the Chester; it is usually laterally compressed. Butts (1941, p. 250) figures two very poorly preserved specimens; Figure 27 is a true *P. maccalliei*. One specimen was collected by the late Dr. Moses N. Elrod, from an unknown locality. The idea that the large size is due to sudden change to silty environment (Allen and Lester, 1953, p. 195) may be strongly questioned.

Plesiotype: Ind. Univ. Paleo.Coll.; No. 4612, locality and horizon unknown.

## Pentremites obesus Lyon

### (Pl. 8, figs. 4–6)

*Pentremites obesus* LYON, 1857, Geol. Surv. Ky., vol. 3, p. 469, Pl. 2, figs. 1–1d (Golconda formation, Crittenden Co., Ky).—HALL, 1858, Rept. Geol. Surv. Iowa, vol. 1, pt. 2, p. 695, Pl. 25, fig. 15 (Chester limestone, Southern Illinois and Kentucky).—LESLEY, 1889, Geol. Surv. Penna., Rept. P. 4, vol. 2, p. 620, Fig. p. 621 (Reproduction of Lyons' original figure).—ULRICH, 1905, U. S. Geol. Surv., Prof. Paper 36, p. 64, Pl. 7, figs. 1–4 (Crittenden Co., Ky.).—BUTTS, 1917, Miss. Form. West Va., Ky. Geol. Surv., p. 95, Pl. 23, figs. 7–9 (Figs. after Ulrich, 1905).—WELLER, 1920, Ill. Geol. Surv., Bull. 41, p. 355, Pl. 10, figs. 1–3 (Lower Golconda limestone, Pope Co., Ill.)
*Pentremites spicatus altipelvis* HAAS, 1945, Amer. Mus. Novitates, 1289, p. 2, Figs. 3, 4, 6 (not 11, 12, 13, which are *P. halli*) (Glen Dean limestone, Crane, Martin Co., Ind.)

Calyx ovoid, very large; length up to 63 mm; width up to 51 mm, greatest width submedian; L/W, 1.2; vault subglobular, with strongly convex sides and narrow shallowly concave summit; length 38 mm; pelvis 25 mm long; V/P, 1.5; sides of pelvis slightly concave; pelvic angle 90°–95°; ambulacra broadly V-shaped, both sides of ambulacrum convex to median groove; ambulacral rims only moderately high; 7 lateral grooves in 3 mm; interambulacra broadly concave; deltoids up to 19 mm long, projecting 1–3 mm above spiracles.

Common in the Golconda and Glen Dean formations of Indiana, Illinois, and Kentucky.

*Pentremites obesus* has been allowed such a wide latitude in interpretation that forms which have little resemblance to the type have been included with it. Ideas about the allowable variation in a species may change, but the type specimen or figure, if known, remains constant. The usual tendency is to split genera and species into divisions, but the reverse has been true of *P. obesus*.

The description and figures of *P. obesus* Lyon (1857, p. 469) seem distinctive enough so that comparisons between them and unidentified specimens could readily be made. Hall (1858, p. 695) made a correct interpretation of *P. obesus* even though his specimen lacked the slightly protruding deltoids, which are very commonly broken or weathered off. Rowley (1903, p. 115, Pl. 36, figs. 1–3) figured a specimen from Grayson Springs, Ky., as *P. obesus* but it may be *P. spicatus*. Ulrich and Smith (1905, p. 64) illustrated two specimens which show the then generally accepted amount of variation allowed. Weller (1920, Pl. 10, figs. 1–3) also has two widely divergent specimens figured as *P. obesus*. Figures 1–2 may represent *P. obesus*, but Figure 3 is *P. hemisphericus*, unless the restoration of the base is inaccurate.

On the other hand, the species has been called a variety of another species; there is no difference, excepting smaller size, between *P. spicatus altipelvis* Haas and *P. obesus* Lyon.

### Pentremites broadheadi Hambach

(Pl. 9, fig. 1)

*Pentremites broadheadi* HAMBACH, 1880, Trans. St. Louis Acad. Sci., vol. 4, p. 159, Pl. B, fig. 6 (Chester limestone, Evansville, Ill.)

Calyx subovoid, large; length 33 mm; width 31 mm, position of greatest width submedian; L/W, 1.1; sides of vault strongly convex, with concave summit between protruding deltoids; length of vault 22 mm; pelvis 11 mm long; V/P, 2; sides of pelvis sigmoid; basal plates are strongly inflated or nodose; pelvic angle 100°–110°; ambulacra broad, 6–7 mm, and very depressed in center; ambulacral flanges moderately high; transverse grooves or poral pieces, 9 in 0.1 inch, or 10–11 in 3 mm; interambulacra strongly concave; deltoids 7 mm long, extending about 2 mm above spiracles, not flaring.

Chester, probably Glen Dean limestone, Ill.

This species is like *P. spicatus* except for the "very projecting pentangular disc, with five plainly marked surfaces converging at the center." It differs from *P. sulcatus* in the longer deltoids and more inflated basals. *P. broadheadi* is a rare form, for it has not been redescribed, the writers have seen only Hambach's specimens, now in the U. S. National Museum.

### Pentremites halli Galloway and Kaska, n. sp.

(Pl. 9, figs. 2–7; Pl. 13, figs. 4, 5)

*Pentremites cherokeeus* (not Troost, MS., 1849) HALL, 1858, Rept. Geol. Surv. Iowa, vol. 1, pt. 2, p. 691, Pl. 25, fig. 12 a-c (Middle Chester, Chester, Ill.).—ROWLEY *in* GREENE, 1903, Contrib. Ind. Paleo., vol. 1, pt. 12, p. 116, Pl. 36, fig. 3 (Chester, Clifty Station, Ky.).—WELLER, 1920, Ill. Geol. Surv., Bull. 41, p. 371, Pl. 10, figs. 11–13 (Upper Chester, Menard limestone, 5 miles south of Chester, Ill.)
*Pentremites spicatus porrectus* HAAS (part), 1945, Amer. Mus. Novitates, no. 1289, p. 3, Fig. 10 (Glen Dean limestone, Crane, Ind.)
*Pentremites spicatus altipelvis* HAAS (part), 1945, Amer. Mus. Nov. no. 1289, p. 3, Figs. 11–13 (Glen Dean limestone, Crane, Ind.)
*Pentremites cherokeeus* (not Troost, but Hall) HAAS, 1945, Amer. Mus. Novitates, no. 1289, p. 5, Figs. 14–16 (Glen Dean limestone, Crane, Ind.)

Calyx pentagonal in side view, wider than long, up to 25 mm wide and nearly as long; greatest widtd suprabasal, at lower tips of ambulacra; L/W, 0.9–1; sides of vault nearly straight to slightly curved, summit wide and flat, except for protruding deltoids; pelvis of moderate length, 5–7 mm, with nearly straight edges, V/P, 1.4–3; basal angle 110°–130°; ambulacra broad, V-shaped, strongly concave, 5–7 mm wide at deltoids, with sharp flanges up to 1 mm wide; transverse grooves about

9 in 3 mm; interambulacra strongly concave; deltoids 5–7 mm long, extending 1–2 mm above spiracles and flaring outward, transversely thin and fragile, broken off in many specimens.

Middle Chester, Glen Dean, of Chester and Prairie du Long Creek, Ill., Huntsville, Ala., Hardin Co., Ky.; abundant at Cloverport, Ky., and Rome, Ind.; Upper Chester, Menard limestone, Randolph and Johnson cos., Ill.

This species is obviously not *P. cherokeeus* of Troost, which originally came from Cherokee Co., Tenn., first definitely made known by Wood (1909, p. 16, Pl. 3, figs. 14–16), and the type figure of that species. *P. halli* differs from it in not having the deep, V-shaped ambulacra with wide flanges. *P. cherokeeus* is the more advanced form. Hall considered *P. sulcatus* Roemer, 1851, a synonym of *P. cherokeeus* Troost, 1850, but Troost's species was a *nomen nudum* until Wood published the description and figure in 1909. *P. halli* may be only a neanic stage of *P. spicatus*, with which it occurs. It is very similar to Hambach's Figures 5–7 of his ontogenetic series of *P. sulcatus* (1903, Pl. 6) (Pl. 12, figs. 5–7). The pentameral shape, with the five nearly straight sides, and the broad summit and flaring deltoids make *P. halli* easily identifiable, and since it is abundant in the Glen Dean it is a good index fossil for that formation. It is the ancestral species of forms with high deltoids, deep ambulacra with high rims, and flattened bases.

Holotype: Ind. Univ. Paleo. Coll.; No. 5257, Glen Dean limestone, 1 mile east of Cloverport, Ky

### Pentremites cherokeeus Troost

#### (Pl. 9, figs. 8, 9)

*Pentremites cherokeeus* TROOST, 1850, Proc. Amer. Assoc. Adv. Sci., vol. 2, Pl. 60 *Nomen nudum.*—
    WOOD, 1909, U. S. Nat. Mus., Bull. 64, p. 16, Pl. 3, figs. 14–16 (As a synonym of *P. sulcatus* Roemer; gives Troost's original description and figures Troost's holotype) (Base of Lookout Mountain, Cherokee Co., Tenn.)
*Pentremites sulcatus?* ROWLEY, *in* GREENE, 1903, Contrib. to Ind. Paleo., pt. 12, p. 116, Pl. 36, figs. 4–5 (Chester, Big Clifty, Hardin Co., Ky.)

Calyx truncate pyramidal, length and width equal, each 32 mm in type figure; greatest width at base of ambulacra: L/W, 1; vault with slightly sloping and curved sides, length 24 mm; summit nearly flat, about half the width of calyx; pelvis short, 8 mm in the type, with straight sides; V/P, 3–4; pelvic angle 140°; ambulacra V-shaped, very deep and narrow, 4 mm wide, with flanges up to 2 mm wide; transverse grooves small, 9–10 in 3 mm; interambulacra very concave; deltoids about 10 mm long, extending about 2 mm above spiracles, and flaring outward.

Middle Chester, Cherokee Co., Tenn., and Glen Dean formation, 1 mile east of Cloverport, Ky.; Hardin Co., Ky.

This species differs from the less-advanced species, *P. halli*, in the narrower and deeper ambulacra, the narrower summit, and the broader ambulacral flanges. Its other nearest relative appears to be *P. spicatus*, differing from it in the narrow V-shaped ambulacra, shorter and more flaring deltoids.

Plesiotype: Ind. Univ. Paleo. Coll.; No. 5259, Glen Dean limestone, 1 mile east of Cloverport, Ky.

### Pentremites spicatus Ulrich

#### (Pl. 9, figs. 10, 11; Pl. 10, figs. 1–4)

*Pentremites obesus* ROWLEY, *in* GREENE, 1903, Contrib. to Ind. Paleo., pt. 12, p. 115, Pl. 36, figs. 1–3 (Chester limestone, Grayson Springs, Ky.)
*Pentremites spicatus* ULRICH, 1917, Form. Chester Ser. in West. Ky., Ky. Geol. Surv., p. 263, Pl. 7, figs. 33–35 (Cherty beds of Glen Dean limestone, Grayson Co., Ky.).—BUTTS, 1917, Miss. Form. West. Ky., Ky. Geol. Surv., p. 100, Pl. 24, figs. 3–4 (Glen Dean limestone, 1 mile east of Cloverport, Ky.).—WELLER, 1920, Ill. Geol. Surv., Bull. 41, p. 368, Pl. 10, fig. 8 (Upper Okaw limestone, Randolph Co., Ill.; Glen Dean limestone, Pope Co., Ill.; Crittenden and Breckinridge cos., Ky.).—HAAS, 1945, Amer. Mus. Novitates, no. 1289, p. 2, Fig. 8 (Glen Dean limestone, Crane, Martin Co., Ind.)

Calyx ovoid, subglobose to truncate conoidal, very large, length up to 42 mm; width up to 41 mm, greatest width submedian; L/W, 1; vault paraboloid, with strongly convex sides and concave summit, length up to 35 mm; pelvis up to 7 mm long, with concave sides; V/P, 5; ambulacra wide, up to

6 mm, broadly V to broadly U-shaped; pelvic angle 120°–140°; ambulacral flanges very strongly developed, up to 4 mm wide; 8 transverse grooves in 3 mm; interambulacra deeply concave; deltoids up to 18 mm long, extending 4–5 mm above spiracles.

Glen Dean limestone, Grayson, Breckinridge, and Crittenden cos., Ky., Randolph and Pope cos., Ill., and Perry and Martin cos., Ind.

*Pentremites spicatus* is one of the most specialized of the pentremites. The high ambulacral rims, low base, and projecting deltoids make it easy to identify. It is an index fossil of the Glen Dean limestone. The characters of *P. spicatus* have been a subject of controversy for some time. The figure of Hambach (1903, Pl. 6, fig. 12) appears to be typical *P. spicatus* Ulrich. Weller (1920, p. 371) believed that the "*P. sulcatus* Roemer" in Wood (1909, p. 16, Pl. 3, figs. 14–16) was the form to which Ulrich gave the name *P. spicatus*. The writers do not agree with Weller. The figure given by Wood is from the type of *P. cherokeeus* Troost which is easily distinguished from *P. spicatus* Ulrich by the narrow V-shaped ambulacra and flaring deltoids, and from *P. sulcatus* Roemer by the lack of ambulacral flanges and shorter deltoids, and inflated basals. The writers have a specimen identical with Wood's *P. sulcatus*, the type of *P. cherokeeus* Troost, from the Glen Dean limestone at Cloverport, Ky., which is identical with *P. cherokeeus* Troost. Haas (1945, p. 1) believed that the difference between *P. spicatus* and *P. sulcatus* was in the lack of projecting deltoids in the latter. This is true, but the other characteristics that *P. spicatus* possesses which distinguish it from *P. sulcatus* are high ambulacral rims, U-shaped ambulacra, and a concave rather than a sigmoid pelvis. "*P. cherokeeus*" Hall and "*P. cherokeeus*" Weller are the same, but neither is *P. cherokeeus* Troost, first made known by Wood (1909, Pl. 3, figs. 14–16). The writers have named Hall's form *P. halli* new species.

Plesiotypes: Ind. Univ. Paleo. Coll., Nos. 5260, 5261, 5262, Glen Dean limestone, 1 mile east of Cloverport, Ky.

### Pentremites spicatus porrectus Haas

(Pl. 10, figs. 5, 6)

*Pentremites spicatus porrectus* HAAS, 1945, Amer. Mus. Nov., no. 1289, p. 3, Pl. 1, figs. 1, 2, 5, 7, 9, not 10 (Glen Dean limestone, Crane, Martin Co., Ind.)

Calyx ovoid, very large; length from 31–45 mm; width from 25–36 mm; greatest width submedian; L/W, 1.2; vault paraboloid, with moderately convex sides and a moderate to narrow and concave summit; vault length from 25–35 mm; pelvis from 6–11 mm long; V/P, 3.2–4.1, sides concave; pelvic angle 105°–125°; ambulacra deeply concave, with moderately high rims; 7–8 lateral grooves in 3 mm; interambulacra very concave to V-shaped; deltoids 9–13 mm long, projecting 1–2 mm above the spiracles.

Common in the Glen Dean limestone at Crane, Martin Co., Indiana.

*Pentremites spicatus porrectus* is similar to *P. chesterensis* in general shape but has high protruding deltoids, more concave interambulacra, and high ambulacral rims. It occurs with *P. spicatus*, from which it differs in the narrower form and smaller pelvic angle.

Plesiotype: Ind. Univ. Paleo. Coll.; No. 5264, Glen Dean limestone, Crane, Martin Co., Ind.

### Pentremites serratus Hambach

(Pl. 10, figs. 7, 8)

*Pentremites serratus* HAMBACH, 1903, Trans. St. Louis Acad. Sci., vol. 13, p. 56, Pl. 4, fig. 9 (Ste. Genevieve, Mo.; Baldwin, Ill)

Calyx subglobose, very large; length up to 34 mm, width up to 37 mm, greatest width median, L/W, 1.0; vault subround, with strongly convex sides and a small flattened summit, length up to 27 mm; pelvis up to 7 mm long; V/P, 4; sides of pelvis strongly concave, pelvic angle 125°; ambulacra strongly concave, broadly V-shaped with well-developed, strongly serrated ambulacral margins; about 7 transverse grooves in 3 mm; interambulacra deeply concave; deltoids up to 10 mm long, projecting 2–3 mm above spiracles.

Rare from uppermost Middle Chester, Glen Dean formation, in Missouri, Illinois, Kentucky, and Indiana. Found with *P. spicatus* but much less abundant than that species.

This species is distinguished from *Pentremites spicatus* only by the serrate ambulacral rims. It may be only the gerontic stage of *P. spicatus*.

Plesiotype: Ind. Univ. Paleo. Coll.; No. 5263, Glen Dean limestone, 1 mile east of Cloverport, Ky.

## Pentremites basilaris Hambach

### (Pl. 10, figs. 9, 10)

*Pentremites basilaris* HAMBACH, 1880, Trans. St. Louis Acad. Sci., vol. 4, p. 159, Pl. B, fig. 9 (Chester limestone, Evansville, Chester, Ill.)

Calyx subconoidal, very large, length of type 40 mm, width 37 mm; greatest width suprabasal, L/W, 1.1; sides of vault moderately convex, summit narrow, 15 mm, and concave, length up to 35 mm; pelvis nearly flat, length less than 5 mm; V/P, 7; sides a double convex curve; pelvic angle 165°; ambulacra deeply concave with prominent rims 2 mm high; interambulacra deeply concave; about 6 lateral grooves in 3 mm; deltoids very long, up to 15 mm; extending up to 3 mm above the spiracles.

Glen Dean limestone, Evansville, Ill., and Crane, Martin Co., Ind.

This species is distinguished from *P. spicatus* by the flatter base and narrower form.

Plesiotype: Ind. Univ. Paleo. Coll.; No. 5265, Glen Dean limestone, Crane, Martin Co., Ind.

## Pentremites bradleyi Meek

### (Pl. 10, fig. 11)

*Pentremites bradleyi* MEEK, 1873, Geol. Surv. Terrs., 6th Ann. Rep. F. V. Hayden, p. 470 (Not figured) (Carboniferous, Mont.).—HAMBACH, 1903, Trans. St. Louis Acad. Sci., vol. 13, p. 56, Pl. 5, fig. 7 (Meek's specimen, Smithsonian Coll., 24529).—CLARK, 1917, Bull. Mus. Comp. Zool., Harvard, vol. 61, no. 9, p. 368 (No figure: Meek and Hambach's descriptions repeated)

Calyx conoidal, up to 12 mm in length and width, greatest width basal; L/W, 1–1.1; vault conoidal with nearly straight sides and nearly flat summit; pelvis short, 2–3 mm, with nearly straight sides; V/P, 3–5; pelvic angle 135°–145°; ambulacra narrow, V-shaped, with narrow, very slightly raised rim; transverse grooves 9 in 3 mm; interambulacra slightly concave; deltoids short, 2–3 mm, not reaching summit.

At least eight specimens from the Golconda formation, from the bluff of Beaver Creek half a mile southeast of Shoals, Ind., fit the descriptions and figures of this species; they tend to have a length to width ratio of 1.1 instead of 1. This species differs from *P. tulipaformis* in the larger basal angle, lack of basal nodes, and conoidal form. The concave instead of convex ambulacra immediately distinguish it from *P. conoideus* and from *P. godoni*.

## Pentremites globosus Say, 1825

"1. *P. globosa*. Body subglobular; sutures with parallel impressed lines. Length one inch and one-fifth; greatest breadth one inch and three-tenths... This large and fine species belongs to the Philadelphia Museum. It was brought from England by Mr. Reubens Peale, who understood that it was found in the vicinity of Bath." SAY, Jour. Acad. Nat. Sci. Philadelphia, ser. 1, vol. 4, 1825, p. 293. Reprint, Bull. Am. Pal., vol. 1, no. 5, 1896, p. 81. Never redescribed or figured, and unrecognizable.

"1. *Pentremites globosa*, Say, supposed to have been found at Bath in England. We have found it in Alabama, at Mount Sano and vicinity; Tennessee, Crab Orchard mountain; and Illinois." TROOST, 1835, Trans. Geol. Soc. Penn., vol. 1, pt. 2, p. 228. *P. globosus* is Say's species, not Troost's, and is unrecognizable.

*Pentremites globosus* "Troost, MS. of Monograph", HALL, 1858, Rept. Geol. Surv. Iowa, vol. 1, pt. 2, p. 695, Pl. 25, fig. 17 (Kaskaskia limestone, Hardin Co., Ill.). Small, globular form, 8 mm in height and width, different from Say's species, not to be attributed to Troost, and probably unrecognizable. It may well be, as Hall says (p. 696), "the young of *P. sulcatus*."

*Pentremites globosus* Say in Troost's MSS., 1850; published by WOOD, U. S. Nat. Mus. Bull. No. 64, 1909, p. 13, Pl. 3, fig. 5. Reported by Troost (Wood, p. 13) from rocks of Chester age from several States, and assigned by Wood to *Pentremites godoni*. It is *P. godoni abbreviatus* Hambach.

Etheridge and Carpenter say (1886, p. 158), "So much difference of opinion, however, appears to exist as to what is *P. globosa*, Say, that we think it would much simplify matters were this name totally expunged . . . it is not improbably our *Acentrotremites ellipticus.*"

Neanic stages of most species tend to be globular, and the species to which they belong can be determined only by their association and construction of ontogenetic series. In view of the confusion attending the name *P. globosus* it is best not to use the name for any species of *Pentremites*.

### Pentremites grandis Warren

*Pentremites grandis* WARREN, 1927, Canada Geol. Surv., Mem. 153, p. 48, Pl. 3, fig. 8 (Mississippian, Pennsylvanian?, Rundle limestone, Cascade Mountain, Alberta)

Specimen imperfect, 50 mm long, 32 mm wide; deltoids 27 mm long; base unknown; transverse ridges about 14 in 5 mm.

### Pentremites perelongatus Warren

*Pentremites perelongatus* WARREN, 1927, Canada Geol. Surv., Mem. 153, p. 48, Pl. 3, fig. 9 (Mississippian, Pennsylvanian?, Rundle limestone, Squaw Mountain, Alberta)

Specimen fragmentary, 54 mm long, 25 mm wide, deltoids small, 12 mm long; base unknown; transverse ridges about 12 in 5 mm.

## COLLECTING LOCALITIES

Some localities where abundant specimens of *Pentremites* occur are:

(1) 1 mile south of Harrodsburg, Ind., on Hwy. 37, top of the Harrodsburg and base of the Salem limestones. *P. conoideus* and its varieties.

(2) Quarry half a mile east of Silverville, Lawrence Co., Ind., Beaver Bend limestone, *P. godoni*, *P. malotti*, *P. princetonensis*, and others.

(3) Prairie du Long Creek, east of Vogel School, a quarter of a mile northwest of Floraville, St. Clair Co., Ill., Paint Creek formation. *P. godoni*, *P. symmetricus*, *P. pyramidatus*.

(4) Downeys Bluff, Rosiclare, Ill., Renault formation. *P. godoni*, *P. princetonensis*, *P. pulchellus*.

(5) Lusk Creek, Golconda, Ill., Golconda formation. *P. elegans*, *P. springeri*, *P. godoni*, *P. pyriformis*, *P. welleri*, *P. platybasis*, *P. okawensis*, *P. kirki*, *P. hambachi*.

(6) Beaver Creek, half a mile southeast of Shoals, Ind., Golconda formation. *P. pyriformis*, *P. elegans*, *P. girtyi*, *P. okawensis*, *P. clavatus*, *P. platybasis*, *P. springeri*, *P. welleri*.

(7) 2 miles east of Sulphur, Crawford Co., Ind., Golconda formation. *P. okawensis*, *P. pyriformis*, *p. patei*, *P. welleri*, *P. springeri*, *P. elegans*.

(8) Highway 37, north and south of Grantsburg, Crawford Co., Ind., Golconda formation. *P. springeri*, *P. pyriformis*, *p. girtyi*, *P. okawensis*, *P. patei*, *P. elegans*.

(9) 1 mile south of Gerald, Perry Co., Ind., Golconda formation. *P. tulipaformis*.

(10) Claymore, Elkton, Ky., Golconda formation. *P. pyriformis*, *P. elegans*, *P. springeri*, *P. okawensis*.

(11) Sloans Valley, Ky., Middle Chester. *P. elegans*, *P. pyriformis*, *P. fohsi*, *P. tulipaformis*.

(12) Quarries at Rome, Ind., and Cloverport, Ky., Glen Dean formation. *P. spicatus*, *P. halli*, *P. serratus*, *P. elegans*, *P. patei*, *P. pyramidatus*, *P. platybasis*, *P. symmetricus*, *P. tulipaformis*, *P. cherokeeus*.

(13) Crane Naval Depot, Martin Co., Glen Dean formation. *P. spicatus* and varieties, *P. halli*, *P. tulipaformis*, *P. okawensis*, *P. pyriformis*.

(14) Quarry 2 miles southeast of Herod, Ill., Glen Dean formation. *P. halli*.

(15) Menard, Ill., Menard limestone. *P. sulcatus*.

(16) 8 miles northeast of Marion, Ky., Kinkaid formation. *P. tulipaformis*, *P. gutschicki*, *P. pyramidatus*.

(17) 2 miles west of Robbs, Ill., Kinkaid formation. *P. tulipaformis*.

(18) Sawney Hollow, Okla., and Washington Co., Ark., Brentwood limestone. *P. rusticus*, *P. angustus*. (Mather, 1915)

Most of the localities cited by Lyon (1857; 1860), Hambach (1880; 1884; 1903), Ulrich and Smith (1905), Ulrich (1917), Rowley (1901–1904, 1905), and even by Weller (1920) are not precise enough to indicate the actual spot where the pentremites were found—nor are some of those listed here.

Once the fossils of a given locality are thoroughly collected, only few specimens may be collected later. This is especially true where the rock is limestone. It is scarcely possible to find a pentremite at the localities north and south of Grantsburg, Indiana, along Highway 37, where Dr. Malott collected some 2000 specimens. Well-preserved specimens may be collected year after year along the creek northwest of Floraville, Illinois.

## CHECK LIST OF SPECIES OF *PENTREMITES*

Synonyms, unrecognizable forms, and forms belonging to other genera are in italics.

*P. abbreviatus* Hambach, 1880 = P. godoni abbreviatus Hambach, 1880
P. abruptus Ulrich, 1917
*P. abruptus* var. Ulrich, 1917 = young P. welleri Ulrich, 1917
*P. altimarginatus* Clark, 1917 = P. biconvexus Ulrich, 1917
P. altus Rowley, 1901
P. angularis Lyon, 1860
*P. angustus* Hambach, 1903 = P. godoni angustus Hambach, 1903
*P. arctibrachiatus* Ulrich, 1917 = young P. welleri Ulrich, 1917
*P. arctibrachiatus huntsvillensis* Ulrich, 1917 = young P. welleri Ulrich, 1917
P. basilaris Hambach, 1880
*P. benedicti* Rowley, 1900 = young P. conoideus Hall, 1858
P. biconvexus Ulrich, 1917
*P. bipyramidalis* Hall, 1858 = Metabalastus
P. bradleyi Meek, 1873
*P. brevis* Ulrich 1917 = Pentremites godoni abbreviatus Hambach, 1886
P. broadheadi Hambach, 1880
P. burlingtonensis Meek and Worthen, 1870
P. buttsi Ulrich, 1917
*P. calyce* Hall, 1862, unrecognizable species of Pentremitidea
*P. calycinus* Lyon, 1860 = young P. elegans Lyon, 1860
*P. canalis* Ulrich, 1917 = P. elegans Lyon, 1860
*P. canalis praeciptus* Ulrich, 1917, p. 147. *Nomen nudum*
*P. cariodes* Owen, 1843. *Nomen nudum*
*P. cavus* Ulrich, 1905 = P. conoideus amplus Rowley, 1902
P. cervinus Hall, 1858
*P. cherokeeus* Troost, 1850, 1909
*P. cherokeeus* Hall, 1858, and Weller, 1920 = P. halli n. sp.
P. chesterensis Hambach, 1880
P. clavatus Hambach, 1880
P. conoideus Hall, 1856, 1858
P. conoideus amplus Rowley, 1902
P. conoideus obtusus Hambach, 1903
P. conoideus perlongus Rowley, 1902
*P. conoideus* var. ?, Butts, 1926
*P. cornutus* Meek & Worthen, 1861 = Heteroblastus
*P. curtus* Shumard, 1855 = Orbitremites

*P. decipiens* Ulrich, 1917 = immature stages of P. symmetricus Hall, 1858

*P. decipiens decurtatus* Ulrich, 1917 = P. symmetricus Hall, 1858

*P. decussatus* Shumard, 1858, fragment

P. decussatus Shumard, Weller, 1909

*P. divergens* Clark, 1917 = P. princetonensis Ulrich, 1917

*P. downeyensis* Ulrich, 1917, p. 170, probably P. conoideus Hall 1856

P. elegans Lyon, 1860

P. elongatus Shumard, 1855

*P. florealis* Say, 1825 = P. godoni (Defrance), 1819

P. fohsi Ulrich, 1905

*P. fohsi marionensis* Ulrich, 1905 = P. robustus Lyon, 1860

P. gemmiformis Hambach, 1884, probably young of P. welleri Ulrich, 1917

P. girtyi Ulrich, 1917

*P. globosa* Say, 1825, unrecognizable, possibly Acentrotremites ellipticus (Cumberland)

*P. globosa* Say *in* Troost, 1835, unrecognizable

*P. globosus* Troost, Hall, 1858, unrecognizable young stage; neanic stages of most species globose

*P. globosus* Troost, *in* Wood, 1909 = P. godoni abbreviatus Hambach

P. godoni (Defrance), 1819

P. godoni abbreviatus Hambach, 1886

P. godoni angustus Hambach, 1903

P. godoni major Etheridge and Carpenter, 1886

P. godoni pinguis Ulrich, 1917

P. grandis Warren, 1927

*P. (Granatocrinus) granulosus* Meek & Worthen, 1865 = Cryptoblastus granulosus (M. & W.), 1865

*P. grosvenori* Shumard, 1858 = Troostocrinus

P. gutschicki n. sp.

P. halli n. sp.

*P. hambachi* Ulrich, 1917, p. 226. *Nomen nudum*

P. hambachi Butts, 1926

P. hemisphericus Hambach, 1880

*P. (Codaster) kentuckyensis* Shumard, 1858, unrecognizable

P. kirki Hambach, 1903

*P. koninckana* Hall, 1856 = neanic stage of P. conoideus Hall, 1858

P. laminatus Easton, 1943

*P. laterniformis* Owen and Shumard, 1850 = Orophocrinus or unrecognizable

*P. leda* Hall, 1862 = Pentremitidea leda (Hall); Devonoblastus Reimann, 1935

*P. lineatus* Shumard, 1858 = Metablastus

*P. longicostalis* Hall, 1860, not recognized, or Metablastus

*P. lycorias* Hall, 1862, not definitely recognizable; Pentremitidea

*P. lyoni* Ulrich, 1917 = small adult P. welleri Ulrich, 1917

*P. lyoni gracilens* Ulrich, 1917 = P. kirki Hambach, 1903

*P. maia* Hall, 1862 = Pentremitidea

P. maccalliei Schuchert, 1906

P. malotti n. sp.

*P. melo* Owen and Shumard, 1850 = Cryptoblastus

*P. melo* var. *projectus* Meek and Worthen, 1861 = Cryptoblastus

*P. missouriensis* Swallow, 1860, not recognized

P. nodosus Hambach, 1880

*P. norwoodi* Owen and Shumard, 1850 = Orbitremites

P. obesus Lyon, 1857

*P. obesus modestus* Ulrich, 1917, p. 226. *Nomen nudum*

*P. obliquatus* Roemer, Stafford, 1869 = Tricoelocrinus

*P. obtusus* Hambach, 1903 = P. conoideus obtusus Hambach

P. okawensis Weller, 1920
P. ovalis (Goldfuss), 1826
P. ovoides Ulrich, 1917
P. patei Ulrich, 1917
*P. pediculatus* Ulrich, 1917 = neanic P. welleri Ulrich, 1917
P. perelongatus Warren, 1927
*P. pinguis* Ulrich, 1917 = P. godoni pinguis Ulrich, 1917
*P. planus* Ulrich, 1917 = P. godoni (Defrance), 1819
*P. planus* var. *transitus* Ulrich, 1917, p. 147. *Nomen nudum*
*P. planus* var. Ulrich, 1917 = varieties of P. godoni (Defrance)
P. platybasis Weller, 1920
*P. potteri* Hambach, 1880 = Schizoblastus
*P. prematurus* Ulrich, 1917 = young P. elegans Lyon, 1860
P. princetonensis Ulrich, 1917
P. pulchellus Ulrich, 1917
P. pyramidatus Ulrich, 1905
*P. pyramidatus planulatus* Ulrich, 1917, p. 226. *Nomen nudum*
P. pyriformis Say, 1825
*P. pyriformis constrictus* Ulrich, 1917, p. 147. *Nomen nudum*
P. robustus Lyon, 1860
*P. reinwardtii* Troost, 1835 = Troostocrinus reinwardti (Troost)
*P. roemeri* Shumard, 1855 = Schizoblastus
P. rusticus Hambach, 1903
*P. sampsoni* Hambach, 1884 = Schizoblastus
*P. saxiomontanus* Clark, 1917 = P. symmetricus Hall, 1858
*P. sayi* Shumard, 1855 = Schizoblastus
P. serratus Hambach, 1903
*P. simulans* Ulrich, 1917, p. 226. *Nomen nudum*
*P. sirius* White, 1862 = Orophocrinus
P. speciosus Rowley, 1903
P. spicatus Ulrich, 1917
*P. spicatus altipelvis* Haas, 1945 = P. obesus Lyon, 1860
P. spicatus porrectus Haas, 1945
P. spinosus Hambach, 1880
P. springeri Ulrich, 1917
*P. stelliformis* Owen and Shumard, 1850 = Orophocrinus
*P. subconoideus* Meek, 1873, not recognized
*P. subcylindrica* Hall, and Whitfield, 1875 = Troostocrinus
*P. subplanus* Ulrich, 1917, p. 226. *Nomen nudum*
*P. subspinosus* Greger, 1934, in error for P. spinosus Hambach, 1880
*P. subtruncatus* Hall, 1858 = Codaster
P. sulcatus (Roemer), 1851
P. symmetricus Hall, 1858
*P. tennesseeae* Troost, 1850, not recognizable
*P. transitus* Ulrich, 1917, p. 185. *Nomen nudum*
*P. truncata* Conrad, 1843, not recognized
P. tulipaformis Hambach, 1903
P. turbinatus Hambach, 1903 = young stage of some species in the P. pyriformis group
P. tuscumbiae Ulrich, 1917
*P. (Tricoelocrinus) varsouviensis* Worten, 1875 = Metablastus
*P. verneulli* Troost, 1841 = Nucleocrinus
P. welleri Ulrich, 1917
*P. whitei* Hall, 1862, not definitely recognizable; Pentremitidea
*P. (Troostocrinus) woodmani* Meek and Worthen, 1868 = Tricoelocrinus

*P. wortheni* Hall, 1858 = Metablastus
*P. (species undetermined)* Rogers, 1868 = P. godoni (Defrance), 1819
*P. sp.* Simonds, 1888 = P. godoni angustus Hambach, 1903
*P. sp.?* Rowley, 1903 = P. speciosus Rowley, 1903
*Encrina Godonii* Defrance, 1819 = P. godoni (Defrance), 1819
*Encrinites florealis* Schlotheim, 1820 = P. godoni (Defrance), 1819
*Pentatremites ovalis* Goldfuss, 1826 = Pentremites ovalis (Goldfuss)
*Pentatremites florealis* Goldfuss, 1826 = Pentremites godoni (Defrance), 1819
*Pentatremites granulatus* Roemer, 1851 = Cryptoblastus granulosus (Meek and Worthen), 1865
*Pentatrematites florealis* Roemer, 1851 = Pentremites godoni (Defrance), 1819
*Pentatrematites sulcatus* Roemer, 1851 = Pentremites sulcatus (Roemer), 1851
*Pentatrematites* Steininger, 1853 = Pentremitidea

# BIBLIOGRAPHY

ALLEN, A. T., AND LESTER, J. G. (1953) *Ecological Significance of a Mississippian Blastoid*, Ga. Geol. Surv. Bull. 60, p. 190–199, Pls. 1–6

—— (1954) *Contributions to the Paleontology of Northwest Georgia*, Ga. Geol. Surv. Bull. 62, Pl. 27, fig. 15; Pl. 29, figs. 1–29; Pl. 33, figs. 9, 10

BASSLER, R. S. (1938) *Animalia, Pelmatozoa, Paleozoica*, Fossilium Catalogus, W. Junk, Berlin, pt. 83, p. 1–94

BILLINGS, E. (1869–1870) *Notes on the Structure of the Blastoidea*, Am. Jour. Sci., 2nd ser., vol. 47, p. 353

—— (1869–1870) *Notes on the Structure of the Crinoidea, Cystidea, and Blastoidea*, Amer. Jour. Sci., 2nd ser., 1869, vol. 48, p. 69–83, Fig. 12; 1870, vol. 49, p. 51–58; 1870, vol. 50, p. 225–240

BEEDE, J. W. (1906) *Fauna of the Salem Limestone in Indiana*, Ind. Dept. Geol. and Nat. Res., 30th Ann. Rep., 1905, p. 1263, Pl. 7, figs. 7, 8; Pl. 26, figs. 32, 33

BRONN, H. G., AND ROEMER, F. (1851–1856) *Bronn's Lethaea Geognostica*, Stuttgart, p. 278–283, Pl. 4, figs. 8, 12

BURMA, B. H. (1949) *Multivariate Analysis—A New Analytical Tool for Paleontology and Geology*, Jour. Paleo., vol. 23, p. 95–103

BUTTS, C. (1917) *Mississippian Formations of Western Kentucky*, Ky. Geol. Surv., p. 59, 61, 78, 95, 99, Pls. 14, 15, 21, 23, 24

—— (1926) *Geology of Alabama. The Paleozoic Rocks*, Ala. Geol. Surv., Spec. Rept. 14, p. 180, Pl. 59, figs, 1, 2, 4, 7–9, 10, 11; p. 198, Pl. 65, figs. 1–4

—— (1941) *Geology of the Appalachian Valley in Virginia. Part 2, Fossil Plates and Explanations*, Va. Geol. Surv. Bull. 52, p. 249–250, Pl. 132, figs. 14–29

CARPENTER, P. H. (1881) *On Certain Points in the Morphology of the Blastoidea*, Ann. Mag. Nat. Hist., 5th ser., vol. 8, p. 418

CHRISTY, D. (1848) *Letters on Geology*, J. M. Christy, Rossville, Pl. 4, figs. 6–8

CLARK, T. H. (1917) *New Blastoids and Brachiopods from the Rocky Mountains*, Bull. Mus. Comp. Zool. at Harvard Coll., vol. 61, no. 9, p. 361–371, Pl. 1, figs. 1–14

CONRAD, T. A. (1843) *Observations on the lead-bearing limestone of Wisconsin and descriptions of a new genus of trilobites and fifteen new Silurian fossils*, Proc., Acad. Nat. Sci. Philadelphia, p. 334

COX, E. T. (1871) *Martin County*, 2nd Rept., Geol. Surv. Ind., 1870, p. 81

CRONEIS, C., AND GEIS, H. L. (1940) *Microscopic Pelmatozoa; Part 1, Ontogeny of the Blastoidea*, Jour. Paleo., vol. 14, p. 345–354, Figs. 1–4

CUVIER, GEORGES (1818) *Theory of the Earth*, New York, Kirk & Mercein, p. 363

DEFRANCE, J. L. M. (1819) *Dictionnaire des Sciences Naturelles*, Strasbourg, F. G. Levrault; Paris, Le Normant, vol. 14, p. 467

EASTON, W. H. (1943) *The Fauna of the Pitkin Formation of Arkansas*, Jour. Paleo., vol. 17, p. 125–154, Pl. 21, fig. 10

ETHERIDGE, R. JR., AND CARPENTER, P. H. (1886) *Catalogue of the Blastoidea in the Geological Department of the British Museum (Natural History)*, London, p. 1–310, 20 pls.

GOLDFUSS, A. (1826) *Petrefacta Germaniae*, List & Franke, Leipzig, pt. 1, p. 150, Pl. 50, figs. 1 a, b, c

GRABAU, A. W., AND SHIMER, H. W. (1910) *North American Index Fossils*, A. G. Seiler and Co., New York, vol 2, p. 481–483, Figs. 1792–1794

GREGER, D. K. (1934) *Bibliographic Index of North American Species of the Eublastoidea*, Trans. St. Louis Acad. Sci., vol. 28, no. 3–4, p. 146–177

HAAS, O. (1945) *Remarks on Some Chester Pentremites*, Amer. Mus. Nov., no. 1289, p. 1–9, Figs. 1–42

—— (1946) *Annotated Faunal List of the Glen Dean Formation of Crane, Indiana*, Amer. Mus. Nov., no. 1307, p. 1

HALL, J. (1856) *Descriptions of new Species of Fossils from the Carboniferous Limestone of Indiana and Illinois*, Trans., Albany Inst. vol. 4, p. 4

—— (1858) *Paleontology of Iowa*, Rept. Geol. Surv. Iowa., vol. 1, pt. 2, p. 484–485, 607, 656, 690–696, Pls. 15, 25

—— (1860) *Paleontology of Iowa*, Supplement 1, pt. 2, Geol. Rept. Iowa, Pl. 1, fig. 2 a–c

—— (1862) *Contributions to Paleontology*, State Cabinet of Nat. Hist. N. Y., 15th Ann. Rept., p. 149–151, Pl. 1, figs. 10, 11

—— (1883) *Paleontology*, 12th Ann. Rept. Dept. Geol. Nat. Hist. for 1882, p. 239–375, p. 32, fig. 32

HAMBACH, G. (1880) *Contribution to the Anatomy of the Genus* Pentremites, *with Description of New Species*, Trans. St. Louis Acad. Sci., vol. 4, p. 145–160, Pls. A–B

—— (1884a) *Notes About the Structure and Classification of the Genus* Pentremites, Trans. St. Louis Acad. Sci., vol. 4, p. 537–547, Figs. 1–5

—— (1884b) *Description of New Paleozoic Echinodermata*, Trans. St. Louis Acad. Sci., vol. 4, p. 548–554, Pl. D, figs. 2–7

—— (1903) *A Revision of the Blastoidea with a Proposed New Classification and Description of New Species*, Trans. St. Louis Acad. Sci., vol. 13, p. 1–67, 6 Pls.

HAYDEN, F. V. (1873) Pentremites subconoideus *Meek*, U. S. Geol. Surv. Terrs., 6th Ann. Rept., p. 471

LESLEY, J. P. (1889) *A Dictionary of the Fossils of Pennsylvania and Neighboring States*, Geol. Surv. Penna., Rept. P. 4, vol. 2, p. 619–622, Figs. XI, XII

KEYES, C. R. (1894) *Paleontology of Missouri*, Geol. Surv. Mo., vol. 4, pt. 1, p. 133–136, Pl. 18

LYON, S. S. (1857) *Paleontological Report*, Geol. Surv. Ky., vol. 3, p. 467–472, Pl. 2, figs. 1, 1a–1d

—— (1860) *Descriptions of Four New Species of Blastoidea from the Subcarboniferous Rocks of Kentucky*, Trans. St. Louis Acad. Sci., vol. 1, no. 4, p. 628–634, Pl. 20

LYON, S. S., AND CASSEDAY, S. A. (1860) *A Synonymic List of the Echinodermata of the Paleozoic Rocks of North America*, Proc., Amer. Acad. Arts and Sci., vol. 4, p. 296–298

MATHER, K. F. (1915) *The Fauna of the Morrow Group of Arkansas and Oklahoma*, Bull. Denison Univ., vol. 18, p. 100–102, Pl. 3, figs. 10–13

MEEK, F. B. (1873) *Preliminary paleontological report, consisting of lists and descriptions of fossils, with remarks on the ages of the rocks in which they were found, etc.*, U. S. Geol. Surv. Terr. (Hayden), 6th Ann. Rept., p. 470

—— (1874) *Fossils of the Illinois Geological Report*, Am. Jour. Sci., 3rd ser., vol. 7, p. 375–376

MEEK, F. B. AND WORTHEN, A. H. (1862) *Descriptions of new Paleozoic Fossils from Illinois and Iowa*, Proc., Philadelphia Acad. Nat. Sci., 1861, p. 141–142

—— (1865) *Descriptions of New Crinoidea from the Carboniferous Rocks of Illinois*, Proc., Philadelphia Acad. Nat. Sci., p. 165

—— (1868) *Remarks on some types of Carboniferous Crinoidea*, Proc., Philadelphia Acad. Nat. Sci., p. 356

—— (1873) *Paleontology of Illinois*, Geol. Surv. Ill., vol. 5, p. 461

MILLER, S. A. (1889) *North American Geology and Paleontology*, Cincinnati, Ohio, p. 267–268, 3 Figs.

MITCHILL, S. L. (1818) *Observations on the Geology of North America*, p. 363 in Cuvier, Georges, *Theory of the Earth*, New York, Kirk & Mercein, 431 p., 3 Pls.

MOORE, R. C. (1940) *Early Growth Stages of Carboniferous Microcrinoids and Blastoids*, Jour. Paleo. vol. 14, p. 578–583, Fig. 3

MOORE, R. C., LALICKER, C. G., AND FISCHER, A. G. (1952) *Invertebrate Fossils*, McGraw-Hill Book Co., Inc., New York, Toronto, London, p. 594–603, Figs. 17–8, 9, 10

MORGAN, G. D. (1924) *Geology of the Stonewall Quadrangle, Oklahoma*, Bur. Geol., Bull. 2, p. 198, Pl. 33

MORSE, W. C. (1930) *Paleozoic Rocks of Mississippi*, Geol. Surv. Miss., 212 p., Pl. 13

OWEN, D. D. (1842) *Human Footprints in Solid Limestone*, Am. Jour. Sci., vol. 43, p. 20, Fig. 3

OWEN, D. D., AND SHUMARD, B. F. (1850) *Descriptions of fifteen new species of Crinoidea from the Subcarboniferous limetsones of Iowa, Wisconsin, and Minnesota, in the years 1848–49*. Jour. Acad. Nat. Sci. Philadelphia, vol. 2, pt. 2, p. 57–70, Pl. 7

—— (1852) *Descriptions of Seven New Species of Crinoidea from the Subcarboniferous Limestone of Iowa and Illinois*, Jour. Acad. Nat. Sci. Philadelphia, vol. 2, pt. 2, p. 89–94, Pl. 11

—— (1852) *Description of One New Genus and Twenty-two New Species of Crinoidea from the Subcarboniferous Limestone of Iowa*, in Owen, D. D. (1852) Report of a Geological Survey of Wisconsin, Iowa and Minnesota, Philadelphia, Lippincott, Grambo & Co., p. 587–598, Pls. 5A–5B

PARKINSON, J. (1808) *Organic Remains of a Former World*, London, Noraville and Fell, vol. 2, p. 235–236, Pl. 13, figs. 36, 37

PECK, RAYMOND E. (1930) *Blastoids from Brazer Limestones of Utah*, Pan-American Geol., vol. 54, p. 104

PICTET, F. J. (1851) *Traité de Paléontologie*, 2nd ed., Paris, J. B. Baillère, vol. 4, p. 292, Pl. 99, fig. 9

REIMANN, I. G. (1935) *Middle Devonian Blastoids*, Buffalo Soc. Nat. Hist. Bull. 17, p. 23–45

—— (1942) *"Tully" Blastoids in Western New York and Genotype of* Devonblastus, Buffalo Soc. Nat. Hist. Bull. 17, no. 3, p. 46, Pl. 9, fig. 2

ROEMER, F. (1848) *Ueber gegliederte, aus Kalk-Stucken Zusammengesetzte Tentakeln oder Pinnulae auf den sogenannten Ambulacral-Feldern der Pentremiten*, Neues Jahrb. F. Mineral., vol. 16, p. 292–296

—— (1851) *Monographie der fossilen Crinoidenfamilie der Blastoideen und der Gattung Pentatrematites in Besonderen*, Archiv fur Naturgeschichte, Jahrg 17, Bd. 1, p. 323–397, Pls. 4–8

—— (1876) *Lethaea Geognostica*, Lethaea paleozoica, Atlas, Stuttgart, Pl. 41, fig. 1a–c

ROGERS, H. D. (1868) *The Geology of Pennsylvania*, vol. 2, pt. 2, D. Van Nostrand, New York, p. 833, Fig. 688

ROWLEY, R. R. (1900) *New Species of Crinoids, Blastoids and Cystoids from Missouri*, Amer. Geol., vol. 25, p. 69, Figs. 29–32

—— (1901–1904) Pentremites, *in* Greene, G. K., *Contribution to Indiana Paleontology*, Ewing and Zeller, New Albany, Ind., vol. 1, pt. 8, p. 64, Pl. 23; vol. pt. 10, 87–97, Pls. 29–30; vol. pt. 12, p. 115–126, Pl. 36; vol. pt. 19, p. 192–197

—— (1904) *A Review of Dr. G. Hambach's "Revision of the Blastoidea with a Proposed New Classification and Description of New Species"*, *in* Greene, G. K., (1904) *Contribution to Indiana Paleontology*, Ewing and Zeller, New Albany, Ind., vol. 1, pt. 19, p. 192–197

—— (1905) *Missouri Paleontology*, Amer. Geol., vol. 35, no. 4, p. 301–311, Pl. 21, figs. 25, 26

SAFFORD, J. M. (1869) *Geology of Tennessee*, S. C. Mercer, Nashville, p. 359, Fig. 1

SAY, T. (1820a) *Observations on Some Species of Zoophytes, Shells, etc., Principally Fossil*, Amer. Jour. Sci., vol. 2, no. 2, p. 34–45

—— (1820b) *Fossil Zoology* (Reprinted from *The Paleontological Writings of Thomas Say*), Bull. Amer. Paleo., 1896, vol. 1, no. 5, p. 282–293

—— (1825) *On Two Genera and Several Species of Crinoidea*, Jour. Acad. Nat. Sci., Philadelphia, 1st ser., vol. 4, 289–296 (Reprinted from *The Paleontological Writings of Thomas Say*) Bull. Amer. Paleo, 1895, vol. 1, no. 5, p. 347–354 and Zool. Jour. 1826, vol. 2, no. 7, London, p. 311–314

—— (1826) *On Two Genera and Several Species of Crinoidea*, (Reprinted from Jour. Acad. Nat. Sci. Philadelphia, 1825, vol. 4, no. 9) Zool. Jour., vol. 2, London, p. 311–315

VON SCHLOTHEIM, E. F. (1821) *Die Petrefactenkunde*, Gotha, Becker, p. 339

SCHUCHERT, C. (1906) *A New American Pentremite*, Proc., U. S. Nat. Mus., vol. 30, p. 759–760, Figs. 1–3

SHIMER, H. W. AND SHROCK, R. R. (1944) *Index Fossils of North America*, John Wiley and Sons, Inc., New York and London, p. 133–134, Pl. 50

SHROCK, R. R., AND TWENHOFEL, W. H. (1953) *Principles of Invertebrate Paleontology*, McGraw-Hill Book Co., New York, Toronto and London, p. 660–664, Figs. 14–8, 9, 10, 11

SHUMARD, B. F. (1853) *Paleontology*, p. 200 in Marcy, R. B., *Exploration of the Red River of Louisiana in the year 1852*, Washington, D. C., U. S. 32d Cong. 2nd sess., Sen. Ex. Doc. 54

—— (1955) *Paleontology*, 1st and 2nd Ann. Repts., Geol. Surv. Mo., p. 187–188, Pl. B, fig. 4

—— (1858) *Descriptions of New Species of Blastoidea from Paleozoic Rocks of the Western States, with some Observations on the Structure of the Genus* Pentremites, Trans. St. Louis Acad. Sci., vol. 1, no. 2, p. 238–248, Pl. 9

SMITH, E. A. (1906) *Development and Variation of* Pentremites conideus, Ind. Dept. Geol. and Nat. Res., 30th Ann. Rept., 1905, p. 1219–1242, Pls. 43–47

SOWERBY, G. B. (1826) *Note on the foregoing Paper, together with a Description of a new Species of* Pentremites, Zool. Jour. London, vol. 2, no. 7, p. 316–318

SPRINGER, FRANK (1913) *Blastoidea and Crinoidea*, p. 161–172 in *Zittel-Eastman Textbook of Paleontology*, Macmillan, New York, London, vol. 1

STEININGER, J. (1853) *Geognostische Beschreibung der Eifel*, Fr. Lintz'schen, Trier, p. 35

SWALLOW, G. C. (1860) *Descriptions of New Fossils from the Carboniferous and Devonian Rocks of Missouri*, Trans., St. Louis Acad. Sci., vol. 2, p. 81

TROOST, G. (1835) *On the* Pentremites Reinwardtii, *a New Fossil, with Remarks on the Genus* Pentremites *(Say)*, Trans. Geol. Soc. Penna., vol. 1, p. 224–231, Pl. 10, figs. 1–12

—— (1841) Pentremites Reinwardtii, 5th and 6th Repts., Geol. Tenn., p. 14, 58

—— (1850) *A List of the Fossil Crinoids of Tennessee*, Am. Assoc. Adv. Sci., Proc. vol. 2, p. 59–62

TWENHOFEL, W. H., AND SHROCK, R. R. (1935) *Invertebrate Paleontology*, McGraw-Hill Book Co.; New York and London, 1st ed., p. 167–172, 1 fig.

ULRICH, E. O. (1917) *The Formations of the Chester Series in Western Kentucky and Their Correlates Elsewhere*, Ky. Geol. Surv., p. 147, 185, 220, 226; p. 242–248, Pl. 2; p. 253–256, Pl. 5; p. 257–259, Pl. 6; p. 260–263, Pl. 7

ULRICH, E. O., AND SMITH, W. S. T. (1905) *The Lead, Zinc, and Fluorspar Deposits of Western Kentucky*, U. S. Geol. Surv., Prof. Pap. 36, p. 58, 64, Pls. 6, 7

VAN TUYL, F. M. (1925) *Stratigraphy of the Mississippian Formation of Iowa*, Geol. Surv. Iowa, vol. 30, 1921–1922, 1925

WACHSMUTH, CHARLES (1883) *The Structure of the Basal Plates in* Codaster *and* Pentremites, Geol. Surv. Illinois, vol. 7, p. 346–364

—— (1900) *Blastoidea and Crinoidea*, p. 188–198 in *Zittel-Eastman Textbook of Paleontology*, New York, London, vol. 1

WARREN, P. S. (1927) *Banff Area, Alberta*, Canada Geol. Surv. Mem. 153, p. 48, Pl. 3, figs. 8, 9

WELLER, J. MARVIN (1931) *The Mississippian Fauna of Kentucky*, Geol. Surv. Ky., 6th ser., vol. 36, p. 251, Pls. 35–44

WELLER, J. MARVIN, et al. (1948) *Correlation of the Mississippian Formations of North America*, Geol. Soc. Amer. Bull., vol. 59, p. 91–196

WELLER, J. M., AND SUTTON, A. H. (1940) *Mississippian Border of the Eastern Interior Basin*, Am. Assoc. Petrol. Geol. Bull., vol. 24, no. 5, p. 829

WELLER, S. (1898) *Bibliographic Index of North American Carboniferous Invertebrates*, U. S. Geol. Surv. Bull. 153, p. 411–417

—— (1909) *Kinderhook Faunal Studies-V, The Fauna of the Fern Glen Formation*, Geol. Soc. Amer. Bull., vol. 20, p. 288–289, Pl. 11, figs. 28–29

—— (1920) *The Geology of Hardin County*, Ill. State Geol. Surv. Bull. 41, p. 314–326, 355–358, 368–372, Pls. 4, 10

WHITE, C. A. (1863) *Observations on the Summit Structure of* Pentremites . . ., Jour. Boston Soc. Nat. Hist., vol. 7, no. 4, p. 481–489

—— (1881) *Fossils of the Indiana Rocks*, 2nd Ann. Rept., Indiana Dept. Statistics and Geology, 1880, p. 511–512, Pl. 7, fig. 9

WHITFIELD, R. P. (1882) *On the Fauna of the Lower Carboniferous Limestones of Spergen Hill, Indiana*, Am. Mus. Nat. Hist. Bull., vol. 1, no. 3, p. 39–97, Pls. 6–9

—— (1893) *Contributions to the Paleontology of Ohio*, Rept. Geol. Surv. Ohio, vol. 7, p. 466, Pl. 9, fig. 4

WOOD, E. (1909) *A Critical Summary of Troost's Unpublished Manuscript on the Crinoids of Tennessee*, U. S. Nat. Mus. Bull. 64, p. 12–17, Pls. 2–3

WOODS, H. (1946) *Paleontology*, Cambridge Univ. Press, 8th ed. p. 177–184, 3 Figs.

YANDELL, L. P. (1848) *Sur une Pentremite des Etats-Unis*, Soc. Geol. France Bull., tome 5, p. 296

YANDELL, L. P., AND SHUMARD, B. F. (1847) *Contributions to the Geology of Kentucky*, Louisville, 36 p., 1 Pl.

EXPLANATION OF PLATES

PLATE 1.—AMBULACRA OF *PENTREMITES*

All figures × 6

Figure                                                                    Page

1. Middle part of ambulacrum of *P. conoideus* Hall. Ambulacrum narrow, strongly convex; 8 transverse ridges in 3 mm; ridges have short, oblique grooves at outer ends and minute ridges at inner ends and along sides. (*See* Pl. 2, fig. 9.) Upper Harrodsburg limestone, 1 mile northeast of Greenville, Ind. Ind. Univ. Paleo. Coll., No. 5219........................ 10

2. Middle part of ambulacrum of *P. godoni* (Defrance). Ambulacrum nearly flat and wider than *P. conoideus*. Some grooves on transverse ridges extend farther inward than normal; 8 transverse ridges in 3 mm. (*See* Pl. 3, fig. 12.) Paint Creek formation, near Floraville, Ill. Ind. Univ. Paleo. Coll., No. 5226...................................... 10

3. Middle part of ambulacrum of *P. godoni angustus* Hambach. Ambulacrum nearly flat and indistinguishable from *P. godoni*; 8 transverse ridges in 3 mm. (*See* Pl. 13, fig. 1.) Paint Creek formation, near Floraville, Ill. Ind. Univ. Paleo. Coll., No. 5266................. 10

4. Middle part of ambulacrum of *P. symmetricus* Hall. Grooves on transverse ridges very strong; 7 transverse ridges in 3 mm. (*See* Pl. 4, fig. 13.) Paint Creek formation, near Floraville, Ill. Ind. Univ. Paleo. Coll., No. 5235. (*Cf.* Figs. 6, 7; Pl. 13, fig. 7.).............. 10

5. Middle part of ambulacrum of *P. symmetricus* Hall. Well-preserved surface showing fine parallel ridges on transverse ridges, fine ridges at inner ends of transverse ridges, and pits at outer ends of ridges; 7 transverse ridges in 3 mm. Paint Creek formation, near Floraville, Ill. Ind. Univ. Paleo. Coll., No. 5235a.......................................... 10

6. Lower part of ambulacrum of *P. pyramidatus* Ulrich. Grooves on outer ends of transverse ridges extend farther inward toward tip of ambulacrum; 8 transverse ridges in 3 mm. Paint Creek formation, near Floraville, Ill. Ind. Univ. Paleo. Coll., No. 5267a......... 10

7. Middle part of ambulacrum of *P. kirki* Hambach. Transverse ridges end in pits in middle of side plates; 7 ridges in 3 mm. (*See* Pl. 5, fig. 12.) Golconda formation, Lusk Creek, Golconda, Ill. Ind. Univ. Paleo. Coll., No. 5243.................................... 10

8. Middle part of ambulacrum of *P. girtyi* Ulrich. Details much like *P. godoni* and allies; 8 transverse ridges in 3 mm in middle, up to 11 at top of ambulacrum. (*See* Pl. 5, fig. 16.) Golconda formation, Grantsburg, Ind. Ind. Univ. Paleo. Coll., No. 5244............. 10, 59

9. Middle part of ambulacrum of *P. springeri* Ulrich. Pits at junction of lancet plate and side plates, and at outer ends of transverse ridges; hydropores at outer edges; 8–9 transverse ridges in 3 mm. (*See* Pl. 5, fig. 24.) Golconda formation, Lusk Creek, Golconda, Ill. Ind. Univ. Paleo. Coll., No. 5248............................................... 10

10. Middle part of ambulacrum of *P. elegans* Lyon. Grooves on ridges extend nearly to median groove; shows suture between lancet plate and side plates, and pits on side plates; 8 ridges in 3 mm. (*See* Pl. 5, fig. 29.) Glen Dean formation, 1 mile east of Cloverport, Ky. Ind. Univ. Paleo. Coll., No. 5250........................................ 10

11. Middle part of ambulacrum of half-grown specimen of *P. tulipaformis* Hambach. Ambulacrum is concave; 10–11 alternating transverse grooves in 3 mm; lancet plate, suture, side plates, median groove, lateral pores and pits, finer ridges and grooves, and fine ridges swinging around inner ends of ridges, are shown. Kinkaid formation, 2 miles west of Robbs, Ill. Ind. Univ. Paleo. Coll., No. 5268......................................... 10

12. Middle part of ambulacrum of *P. cervinus* Hall. Ambulacrum concave; 9 ridges in 3 mm; grooves at ends of transverse ridges oblique and short; finer structures obscured by silicification. Golconda formation, half a mile north of Grantsburg, Ind. Ind. Univ. Paleo. Coll., No. 5252a................................................................. 10

Hydropores are shown in Figures 6, 7, 8, 9, and 11. Sutures between lancet plate and side plates are shown in Figures 3, 5, 6, and 8–12. Figures have been slightly retouched. *See* also page 17.

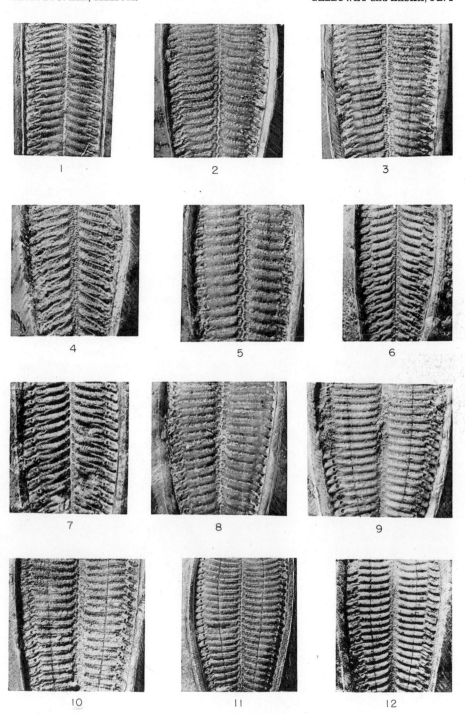

AMBULACRA OF *PENTREMITES*

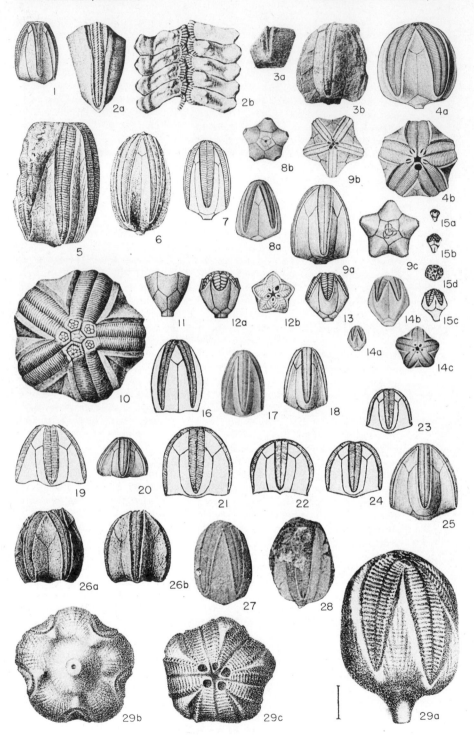

PENTREMITES CONOIDEUS GROUP

PLATE 2.—*PENTREMITES CONOIDEUS* GROUP

All figures × 1 unless otherwise indicated

Figure                                                                                 Page

1. *Pentremitidea leda* Hall (1862, Pl. 1, fig. 11). After type figure. Probable ancestor of *Pentremites*. Side plates cover most of lancet plate. Hamilton shales, western N. Y.. 40

2. *Pentremites decussatus* Shumard (1858, Pl. 9, figs. 6 a, b). After type figure. (a) radial plate and ambulacrum; (b) upper part of ambulacral field. × 6. Lower Misssissippian, Louisville, Ky............ 41

3. *Pentremites decussatus* Shumard. After Weller (1909, Pl. 11, figs. 28, 29). (a) incomplete radial plate, showing ornamentation; (b) nearly complete crushed specimen. Fern Glen, Mo........ 41

4. *Pentremites burlingtonensis* Meek and Worthen (1873, Pl. 8, fig. 7). After type figure. Upper Burlington limestone, Burlington, Iowa.......... 42

5. *Pentremites elongatus* Shumard (1855, Pl. 8, fig. 4). After type figure. Upper Burlington, Mo........ 42

6. *Pentremites elongatus* Shumard (After Keyes, 1894, Pl. 18, fig. 4). Upper Burlington limestone, Mo........ 42

7. *Pentremites elongatus* Shumard (After Etheridge and Carpenter, 1886, Pl. 2, fig. 15). Mississippian, Iowa........ 42

8. *Pentremites conoideus* Hall (1858, Pl. 22, figs. 9, 10, not 8). After type figure. Salem limestone, Spergen Hill or Bloomington, Ind........ 42

9. *Pentremites conoideus* Hall. Upper Harrodsburg limestone, 1 mile northwest of Greenville, Ind. (*See* Pl. 1, fig. 1.) Ind. Univ. Paleo. Coll., No. 5219.......... 42

10. *Pentremites conoideus* Hall (After Shumard, 1858, Pl. 9, fig. 4). Showing plates covering mouth and spiracles. Magnified about 3 times. Salem limestone, Spergen Hill, Ind........ 42

11, 12, 13. *Pentremites conoideus* Hall. Young specimens from the Salem limestone, Old Cleveland Quarry, 1 mile north of Harrodsburg, Ind. (11) nepionic specimen 1 mm long, showing only basals and radials × 12. (12a, b) side and summit view of ananeanic specimen 2 mm long, showing ambulacra and summit structures. × 6. (13) neanic specimen 5 mm long. × 3. Ind. Univ. Paleo. Coll,. No. 1027........ 42

14. *Pentremites conoideus* Hall. Specimen in "*koninckanus*" stage. After type figure of *P. konickanus* Hall (1858, Pl. 22, fig. 11a–c). (a) × 1. (b, c) same specimen, ×2.. 42

15. *Pentremites conoideus* Hall. Young specimens in "*koninckanus*" stage (after type figures of *P.* benedicti Rowley, 1900, Pl. 2, figs. 29–32. (a, b, c) side views of 3 specimens; (d) summit view of another specimen........ 42

16. *Pentremites conoideus perlongus* Rowley (1902, Pl. 29, fig. 28). After type figure. Upper Harrodsburg or Salem limestone, Lanesville, Ind........ 43

17. *Pentremites conoideus perlongus* Rowley (After Hall, 1858, Pl. 22, fig. 8). Hall stated, "...full grown individual, which is more elongate than usual." Salem limestone, Spergen Hill or Bloomington, Ind........ 43

18. *Pentremites conoideus perlongus* Rowley. Upper Harrodsburg limestone, 1½ miles south of Harrodsburg, Ind. Ind. Univ. Paleo. Coll., No. 2392........ 43

19. *Pentremites conoideus obtusus* Hambach (1903, p. 53, Fig. 13). After type figure. Warsaw limestone, Boonville, Mo........ 43

20. *Pentremites conoideus obtusus* Hambach. Upper Harrodsburg limestone, "Pentremite Hollow", 2 miles south of Bloomington, Ind. Ind. Univ. Paleo. Coll., No. 5220.. 43

21–24. *Pentremites conoideus amplus* Rowley (1902, Pl. 29, figs. 31–34). Specimens showing variation in individuals; (24) here designated as lectotype. Upper Harrodsburg or Salem limestone, Lanesville, Ind........ 44

25. *Pentremites conoideus amplus* Rowley. Upper Harrodsburg limestone, 1½ miles south of Harrodsburg, Ind. Ind. Univ. Paleo. Coll., No. 5221......... 44

26. *Pentremites conoideus amplus* Rowley. After type figure of *P. cavus* Ulrich (1905, Pl. 6, figs. 7, 8). Lower St. Louis limestone, Princeton, Ky........ 44

27, 28. *Pentremites ovoides* Ulrich. After type figures (*in* Butts, 1917, Pl. 15, figs. 6, 7.) Ste. Genevieve limestone, Livingston Co., Ky........ 44

29. *Pentremites ovalis* Goldfuss (1862, Pl. 50, fig. 1a, b, c). After type figure. Lack of deltoid plates and doubt whether ambulacra are made up entirely of side plates suggest this form is *Pentremitidea*. Devonian, Germany. Natural size shown by line left of Figure 29a........ 44

87

PLATE 3.—*PENTREMITES GODONI* GROUP

All figures × 1 unless otherwise indicated

Figure                                                                                 Page

1. *Pentremites godoni pinguis* Ulrich (1907, Pl. 2, fig. 17). After type figure. Greater width and shorter base than *P. godoni*. Upper Ste. Genevieve limestone, 2 miles east of Carrsville, Ky. U. S. Nat. Mus. Coll., No. 64796 ................................................. 45

2. *Pentremites godoni pinguis* Ulrich. Paint Creek formation, Prairie du Long Creek, near Floraville, St. Clair Co., Ill. Ind. Univ. Paleo. Coll., No. 5222 ........................... 45

3, 4. *Pentremites tuscumbiae* Ulrich (1917, Pl. 2, figs. 14, 15). After type figures. (3) holotype from the Ste. Genevieve limestone, near Tuscumbia, Ala.; (4) slightly crushed specimen, from the upper Ste. Genevieve limestone, 2 miles east of Carrsville, Ky. Ulrich's figures probably reversed in printing, making (4) the holotype. U. S. Nat. Mus. Coll., Nos. 64795, 64793 ................................................................... 46

5. *Pentremites biconvexus* Ulrich (1917, Pl. 2, figs. 34, 35). After type figure. (a) photograph of holotype; (b) drawing of holotype, with left deltoid too long. Lower Chester limestone, near Bowling Green, Ky. U. S. Nat. Mus. Coll., No. 64820 ................... 47

6. *Pentremites biconvexus* Ulrich. Beaver Bend limestone, old quarry, half a mile east of Silverville, Lawrence Co., Ind. Ind. Univ. Paleo. Coll., No. 5223 ........................ 47

7. *Pentremites biconvexus* Ulrich. After type figure of *P. planus* var. Ulrich (1917. Pl. 5, fig. 14). Paint Creek formation, near Floraville, St. Clair Co., Ill ........................... 47

8. *Pentremites biconvexus* Ulrich. After type figures of *P. altimarginatus* Clark (1917, Pl. 1, figs. 12, 13). (a, b) photograph of silicified type, with poor preservation. × 4/3. (c, d) restored drawing from type figures. × 1. Quadrant limestone, Old Baldy, Mont. Harvard Mus. Comp. Zool. Coll., No. 1003 ................................................. 47

9. *Pentremites malotti* n. sp. Holotype. Beaver Bend limestone, old quarry, half a mile east of Silverville, Lawrence Co., Ind. Ind. Univ. Paleo. Coll., No. 5224 ............... 48

10. *Pentremites malotti* n. sp. Paratype, immature. Beaver Bend limestone, old quarry, half a mile east of Silverville, Lawrence Co., Ind. Ind. Univ. Paleo. Coll., No. 5225 ....... 48

11. *Pentremites godoni* (Defrance). After type figure in Parkinson (1808, Pl. 13, figs. 36, 37). First figures of *Pentremites*, type of genus and type of species. Chester Series, Ky.... 48

12. *Pentremites godoni* (Defrance). Paint Creek formation, Prairie du Long Creek, near Floraville, St. Clair Co., Ill. (*See* Pl. 1, fig. 2.) Ind. Univ. Paleo. Coll., No. 5226 .... 48

13. *Pentremites godoni* (Defrance). After figure of *P. planus* Ulrich (1917, Pl. 5, fig. 6). Paint Creek formation, Prairie du Long Creek, near Floraville, St. Clair Co., Ill ......... 48

14. *Pentremites godoni abbreviatus* Hambach (1886, Pl. B, fig. 3). After type figure. Chester limestone, Evansville, Ill ....................................................................... 49

15. *Pentremites godoni abbreviatus* Hambach. Golconda formation, Golconda, Ill. Ind. Univ. Paleo. Coll., No. 5227 ................................................................... 49

16. *Pentremites godoni abbreviatus* Hambach. After type figure of *P. brevis* Ulrich (*in* Butts, 1917, Pl. 24, fig. 6). Glen Dean limestone, Ky ....................................... 49

17. *Pentremites godoni abbreviatus* Hambach. Drawing from Figure 16 ................... 49

18. *Pentremites rusticus* Hambach (1903, p. 54, Fig. 15). After type figure. Chester limestone, Washington Co., Ark ........................................................................ 50

19. *Pentremites rusticus* Hambach. Lower Pennsylvanian ?, Brentwood limestone ?, probably Arkansas. (a) photograph showing starlike basals; (b) drawing, restored. Ind. Univ. Paleo. Coll., No. 5228 ........................................................................... 50

20. *Pentremites godoni major* Etheridge and Carpenter (1886, Pl. 2, fig. 1). After type figure. Chester Series, Franklin Co., Ky ....................................................... 49

21. *Pentremites godoni major* Etheridge and Carpenter. Paint Creek formation, Prairie du Long Creek, near Floraville, St. Clair Co., Ill. Ind. Univ. Paleo. Coll., No. 5229 .... 49

22. *Pentremites godoni major* Etheridge and Carpenter. After figure of *P. planus* Ulrich (1917, Pl. 5, fig. 15). Lower Chester limestone, near Cowan, Tenn. U. S. Nat. Mus. Coll., No. 64803 ..................................................................................... 49

23. *Pentremites godoni angustus* Hambach (1903, p. 53, Fig. 14). After type figure. Chester Series, Washington Co., Ark ....................................................................... 50

24. *Pentremites godoni angustus* Hambach. Chester Series, Paint Creek, Ill. Ind. Univ. Coll., No. 5230 ......................................................................................... 50

25. *Pentremites platybasis* Weller (1920, Pl. 4, fig. 37). After type figure. Near base of the Golconda formation, three quarters of a mile north of Marigold, Randolph Co., Ill. 50

26–28. *Pentremites platybasis* Weller (1920, pl. 4, figs. 37, 38, 40, 41). Paratypes. Golconda limestone, Randolph and Hardin cos., Ill ............................................................ 50

29. *Pentremites platybasis* Weller. × 1⅓. Beaver Bend limestone, old quarry half a mile east of Silverville, Lawrence Co., Ind. Ind. Univ. Paleo. Co.., No. 5231 ............... 50

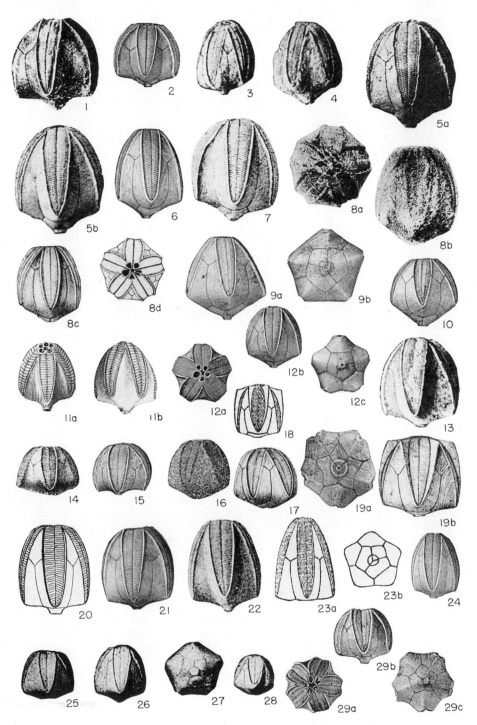

*PENTREMITES GODONI* GROUP

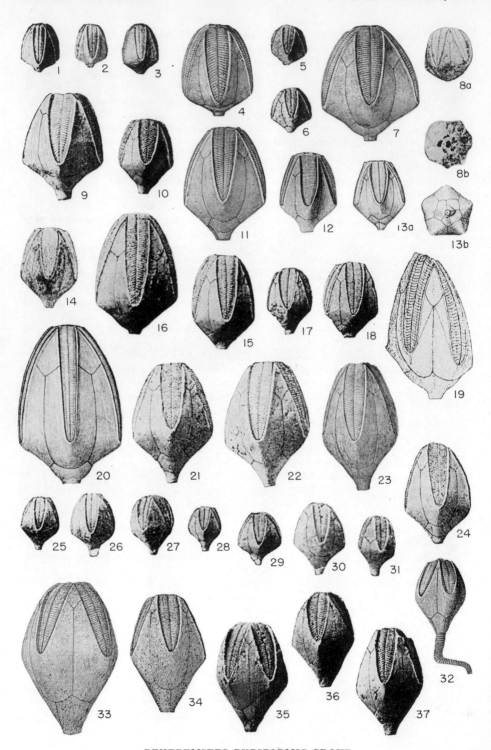

*PENTREMITES PYRIFORMIS* GROUP

PLATE 4.—*PENTREMITES PYRIFORMIS* GROUP

All figures × 1 unless otherwise indicated

Figure                            Page

1, 2. *Pentremites pulchellus* Ulrich (1917, Pl. 2, figs. 2, 3). After type figures. (1) holotype; (2) paratype. Ste. Genevieve limestone, Bakers Station, 2 miles southeast of Fredonia, Ky. U. S. Nat. Mus. Coll., Nos., 64788, 64788a.................................... 51

3. *Pentremites pulchellus* Ulrich. After Weller (1920, Pl. 4, fig. 29). Shetlerville formation, Fairview Bluff, Hardin Co., Ill........................................................ 51

4. *Pentremites pulchellus* Ulrich. × 2. Renault formation, Downeys Bluff, near Rosiclare, Ill. Ind. Univ. Paleo. Coll., No. 5232...................................................... 51

5, 6. *Pentremites princetonensis* Ulrich (1917, Pl. 2, figs. 8, 9). (5) specimen here selected as lectotype; (6) another typical specimen. Both slightly smaller than average size. Cherty zone of Fredonia oölite, Princeton, Ky. U. S. Nat. Mus. Coll., Nos. 64791, 64791a....... 52

7. *Pentremites princetonensis* Ulrich. × 2. Renault formation, Downeys Bluff, near Rosiclare, Ill. Ind. Univ. Paleo. Coll., No. 5233..................................................... 52

8. *Pentremites princetonensis* Ulrich. After type figure of *P. divergens* Clark (1917, pl. 1, figs. 7, 8). Quadrant limestone, Old Baldy, Mont. Harvard Mus. Comp. Zool. Coll., No. 1002................................................................................. 52

9. *Pentremites abruptus* Ulrich (1917, pl. 6, fig. 11). Holotype. Lower Chester limestone, near Cowan, Tenn. U. S. Nat. Mus. Coll., No. 64818............................ 54

10. *Pentremites abruptus* Ulrich (1917, Pl. 6, fig. 14). Paratype. Lower Chester limestone, near Huntsville, Ala. U. S. Nat. Mus. Coll., No. 64819........................... 54

11. *Pentremites abruptus* Ulrich. Glen Dean limestone, old quarries, 1 mile east of Cloverport, Ky. Ind. Univ. Paleo. Coll., No. 5234...................................... 54

12. *Pentremites symmetricus* Hall (1858, Pl. 25, fig. 14). After type figure. Chester limestone, Ky......................................................................... 55

13. *Pentremites symmetricus* Hall. (*See* Pl. 1, fig. 5.) Paint Creek formation, Prairie du Long Creek, near Floraville, St. Clair Co., Ill. Ind. Univ. Paleo. Coll., No. 5235......... 55

14. *Pentremites symmetricus* Hall. After type figure of *P. saxiomontanus* Clark (1917, Pl. 1, fig. 3). Upper Mississippian, Squaw Creek, off Gallatin River, Mont. Harvard Mus. Comp. Zool. Coll., No. 1001.................................................. 55

15, 16. *Pentremites symmetricus* Hall. After Ulrich (1917, Pl. 7, figs. 9, 10). (15) figure Ulrich "proposed to substitute" for Hall's type. Paint Creek formation, Waterloo, Ill. U. S. Nat. Mus. Coll., No. 64814. (16) Renault formation, near Floraville, St. Clair Co., Ill. U. S. Nat. Mus., Springer Coll.................................................. 55

17. *Pentremites symmetricus* Hall. Young specimen. After type figure of *P. decipiens* Ulrich (1917, Pl. 5, fig. 32). Lower Chester limestone, near Cowan, Tenn. U. S. Nat. Mus. Coll., No. 64823a...................................................................... 55

18. *Pentremites symmetricus* Hall. Young specimen. After type figure of *P. decipiens decurtatus* Ulrich (1917, Pl. 5, fig. 34). Lower Chester limestone, near Cowan, Tenn. U. S. Nat. Mus. Coll., No. 64825................................................. 55

19. *Pentremites altus* Rowley (1901, Pl. 23, fig. 1). After type figure. Chester Series, Newmans Ridge, east Tenn. G. K. Greene Coll...................................... 55

20. *Pentremites altus* Rowley. Kinkaid limestone, quarry 2 miles west of Robbs, Ill. Much restored. Coll. Dr. R. C. Gutschick, University of Notre Dame.................... 55

21, 22. *Pentremites welleri* Ulrich (1917, pl. 6, fig. 15). (21) after type figure, here designated as lectotype; (22) paratype. Lower Chester limestone, near Huntsville, Ala. U. S. Nat. Mus., Springer Coll.............................................................. 56

23. *Pentremites welleri* Ulrich. Golconda formation, Lusk Creek, Pope Co., Ill. Ind. Univ. Paleo. Coll., No. 5237.................................................................. 56

24. *Pentremites welleri* Ulrich. After type figure of *P. lyoni* Ulrich (1917, Pl. 7, fig. 27). Golconda formation, Pickering Hill, 7 miles north of Marion, Ky. U. S. Nat. Mus. Coll., No. 64837................................................................. 56

25–27. *Pentremites welleri* Ulrich. Young specimens. After type figures of *P. arctibrachiatus* Ulrich (1917, Pl. 2, figs. 37, 38, 40). Lower Chester limestone, near Huntsville, Ala. U. S. Nat. Mus. Coll., Nos. 64817 a, b........................................... 56

28, 29. *Pentremites welleri* Ulrich. Young specimens. After type figures of *P. arctibrachiatus huntsvillensis* Ulrich (1917, Pl. 2, figs. 41, 42). Lower Chester limestone, Huntsville, Ala. U. S. Nat. Mus., Springer Coll............................................... 56

30. *Pentremites welleri* Ulrich. Young specimen. After type figure of *P. abruptus* var. Ulrich (1917, Pl. 2, fig. 43). Lower Chester limestone, Huntsville, Ala. U. S. Nat. Mus. Coll., No. 64822............ 56

31. *Pentremites welleri* Ulrich. Young specimen. After type figure of *P. pediculatus* Ulrich (1917, Pl. 2, fig. 44). Long stem and long deltoid stated as characteristic. Lower Chester, near Bowling Green, Ky. U. S. Nat. Mus. Coll., No. 64826.............. 56

32. *Pentremites pyriformis* Say. After Hall (1858, Pl. 25, fig. 16). Tapering stem most unusual. Chester limestone, Chester, Ill., Huntsville, Ala............ 56

33. *Pentremites pyriformis* Say. Glen Dean limestone, old quarries 1 mile east of Cloverport, Ky. Ind. Univ. Paleo. Coll., No. 5238............ 56

34. *Pentremites pyriformis* Say. Golconda limestone, half a mile north of Grantsburg, Ind. Ind. Univ. Paleo. Coll., No. 5239............ 56

35–37. *Pentremites pyriformis* Say. After Ulrich (1917, Pl. 6, figs. 5, 6, 9). Lower Chester, Huntsville, Ala., and Bowling Green, Ky............ 56

PLATE 5.—*PENTREMITES PYRIFORMIS* AND *PENTREMITES SULCATUS* GROUPS

All figures × 1 unless otherwise indicated

Figure                                                      Page

1. *Pentremites patei* Ulrich (1917, Pl. 7, fig. 18). After type figure, here designated as lectotype. Lower Chester or Glen Dean limestone, Breckenridge Co., Ky. U. S. Nat. Mus., Springer Coll............ 57

2. *Pentremites patei* Ulrich. Golconda limestone, half a mile north of Grantsburg, Ind. Ind. Univ. Paleo. Coll., No. 5240............ 57

3. *Pentremites pyramidatus* Ulrich (1905, Pl. 7, fig. 13). After type figure, here designated as lectotype. Middle or upper Chester, western Ky. U. S. Nat. Mus., Springer Coll.... 57

4. *Pentremites pyramidatus* Ulrich. Golconda limestone, 1–2 miles south of Herod, Ill. Ind. Univ. Paleo. Coll., No. 5241............ 57

5. *Pentremites gemmiformis* Hambach (1884, Pl. D, fig. 5). After type figure. Lower Chester, probably Paint Creek, Randolph Co., Ill. Washington Univ. Coll............ 58

6. *Pentremites gemmiformis* Hambach. After Ulrich (1917, Pl. 7, fig. 2). Paint Creek formation, Barbeau Creek, Randolph Co., Ill. U. S. Nat. Mus., Springer Coll............ 58

7. *Pentremites buttsi* Ulrich (1917, Pl. 2, fig. 18). After type figure. Lower Chester, Downeys Bluff, Rosiclare, Ill. U. S. Nat. Mus. Coll., No. 64808............ 58

8. *Pentremites buttsi* Ulrich. Paint Creek formation, Prairie du Long Creek, near Floraville, St. Clair, Co., Ill. Ind. Univ. Paleo. Coll., No. 5242............ 58

9. *Pentremites buttsi* Ulrich. After figures of *P. altus* Rowley (1901, Pl. 23, figs. 2, 3). Chester, Bowling Green, Ky. Coll. G. K. Greene............ 58

10. *Pentremites kirki* Hambach (1903, Pl. 5, fig. 18). After type figure. Lower Burlington limestone? Locality not given............ 58

11. *Pentremites kirki* Hambach. After type figure of *P. lyoni gracilens* Ulrich (1917, Pl. 7, fig. 30). Golconda formation, Golconda, Ill. U. S. Nat. Mus. Coll., No. 64840.......... 58

12. *Pentremites kirki* Hambach. (*See* Pl. 1, fig. 7.) Golconda formation, Lusk Creek, Pope Co., Ill. Restored. Ind. Univ. Paleo. Coll., No. 5243............ 58

13. *Pentremites turbinatus* Hambach (1903, Pl. 5, fig. 6). After type figure. Probably middle Chester, Evansville, Ill............ 59

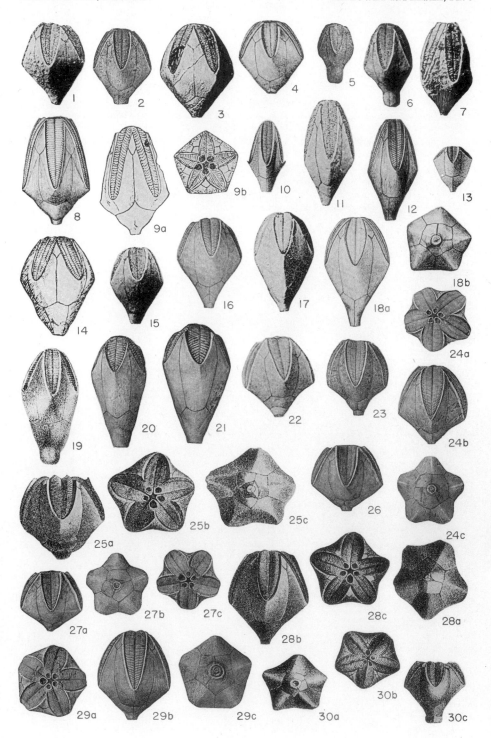

*PENTREMITES PYRIFORMIS* AND *PENTREMITES SULCATUS* GROUPS

PLATE 5 (*Continued*)

14. *Pentremites speciosus* Rowley (1903, Pl. 36, fig. 13). After type figure. Chester, Crittenden Co., Ky. Coll. G. K. Greene.................................................. 59

15. *Pentremites girtyi* Ulrich (1917, Pl. 7, fig. 15). After type figure. Chester, near New Bethel, Breckenridge Co., Ky. U. S. Nat. Mus., Springer Coll............................ 59

16. *Pentremites girtyi* Ulrich. (*See* Pl. 1, fig. 8.) Golconda limestone, half a mile north of Grantsburg, Ind. Ind. Univ. Paleo. Coll., No. 5244............................... 59

17. *Pentremites okawensis* Weller (1920, Pl. 10, fig. 5). After type figure, here selected as lectotype. Golconda limestone, three-quarters of a mile north of Marigold, Randolph Co., Ill................................................................................ 60

18. *Pentremites okawensis* Weller. Golconda limestone, half a mile southeast of Shoals, Ind. Ind. Univ. Paleo. Coll., No. 5245......................................................... 60

19. *Pentremites clavatus* Hambach (1880, Pl. B, fig. 5). After type figure. Chester limestone, Evansville, Ill.................................................................... 60

20, 21. *Pentremites clavatus* Hambach. × 2. Golconda limestone, Beaver Creek, half a mile southeast of Shoals, Ind. Ind. Univ. Paleo. Coll., Nos. 5246, 5246a.................. 60

22. *Pentremites springeri* Ulrich (1917, Pl. 5, fig. 38). After type figure, here designated as lectotype. Lower Chester near New Bethel, Breckenridge Co., Ky. U. S. Nat. Mus., Springer Coll............................................................................ 62

23. *Pentremites springeri* Ulrich. Golconda limestone, half a mile north of Grantsburg, Ind. Ind. Univ. Paleo. Coll., No. 5247......................................................... 62

24. *Pentremites springeri* Ulrich. (*See* Pl. 1, fig. 9.) Golconda formation, Lusk Creek, railroad north of Golconda, Ill. Ind. Univ. Paleo. Coll., No. 5248.................... 62

25. *Pentremites angularis* Lyon (1860, Pl. 20, fig. 3a–c.) After type figure. Glen Dean limestone, Falls of Rough Creek, Breckenridge Co., Ky............................... 64

26, 27. *Pentremites angularis* Lyon. Glen Dean limestone, old quarries, 1 mile east of Cloverport, Ky. Ind. Univ. Paleo. Coll., Nos. 5249, 5249a............................... 64

28. *Pentremites elegans* Lyon (1860, Pl. 20, fig. 4a–c). After type figures. Glen Dean limestone, Grayson Springs, and 3 miles north of Litchfield, Grayson Co., Ky............ 64

29. *Pentremites elegans* Lyon. (*See* Pl. 1, fig. 10.) Glen Dean limestone, old quarries, 1 mile east of Cloverport, Ky. Ind. Univ. Paleo. Coll., No. 5250......................... 64

30. *Pentremites elegans* Lyon. Young specimen. After type figure of *P. calycinus* Lyon (1860, Pl. 20, fig. 1a–c). Golconda limestone, near Grayson Springs and north of Litchfield, Grayson Co., Ky.............................................................. 64

PLATE 6.—*PENTREMITES SULCATUS* GROUP

All figures × 1 unless otherwise indicated

Figure                                                             Page

1, 2, 3. *Pentremites elegans* Lyon. After type figures of *P. canalis* Ulrich (1917, pl. 7, figs. 23, 25, 26). (1a) young ephebic specimen, here selected as lectotype of *P. canalis*: (1b) drawing of (1a); (2) large specimen; (3) summit of another specimen, showing concave ambulacra. Glen Dean limestone, Sloans Valley, Ky. U. S. Nat. Mus. Coll., Nos. 64836, a, b, c.................................................................. 64

4. *Pentremites elegans* Lyon. Neanic stage. After type figure of *P. praematurus* Ulrich (1917, Pl. 2, fig. 20). Ste. Genevieve limestone, Shetlerville, Ill. U. S. Nat. Mus. Coll., No. 64810.................................................................. 64

5. *Pentremites nodosus* Hambach (1880, Pl. B, fig. 2). After type figure. Chester limestone, Randolph Co., Ill.................................................................. 65

6. *Pentremites nodosus* Hambach. After Weller (1920, Pl. 4, fig. 25). Nodes are on radial plates just below deltoids. Golconda limestone, three-quarters of a mile north of Marigold, Randolph Co., Ill.................................................................. 65

7. *Pentremites hambachi* Butts (1926, Pl. 65, fig. 2). After type figure. Bangor limestone, Glen Dean horizon, Frankfort, Franklin Co., Ala. U. S. Nat. Mus., Springer Coll..... 65

8. *Pentremites hambachi* Butts. Golconda formation, Lusk Creek, north of Golconda, Ill. Ind. Univ. Paleo. Coll., No. 5251.................................................................. 65

9. *Pentremites cervinus* Hall (1858, Pl. 25, fig. 11a, b). After type figure. Chester limestone, Chester, Ill., or near Huntsville, Ala.................................................................. 66

10. *Pentremites cervinus* Hall. × 1⅛. Golconda limestone, half a mile north of Grantsburg, Ind. Ind. Univ. Paleo. Coll., No. 5252.................................................................. 66

11, 12. *Pentremites fohsi* Ulrich (1905, Pl. 5, figs. 5–7). After type figures. (11) holotype; (12a, b) base and summit of another specimen. Middle or upper Chester, 8 miles, northwest of Princeton, Ky.................................................................. 66

13. *Pentremites fohsi* Ulrich. Glen Dean limestone, Sloans Valley, Ky. Ind. Univ. Paleo. Coll., No. 5253.................................................................. 66

14. *Pentremites robustus* Lyon (1860, Pl. 20, fig. 2a–c). After type figure. Glen Dean limestone, Grayson Co., Ky.................................................................. 66

15. *Pentremites robustus* Lyon. After type figure of *P. fohsi marionensis* Ulrich (1905, Pl. 7, figs. 10, 11). Middle or upper Chester, railroad cut near Marion, Ky.................................................................. 66

16. *Pentremites tulipaformis* Hambach. (1903, Pl. 4, figs. 10, 11). After type figure. Chester limestone, Kaskaskia, Ill.................................................................. 67

17. *Pentremites tulipaformis* Hambach. Kinkaid limestone, Alexander Stone Co. quarry, 8 miles northwest of Marion, Ky. Ind. Univ. Paleo. Coll., No. 5254.................................................................. 67

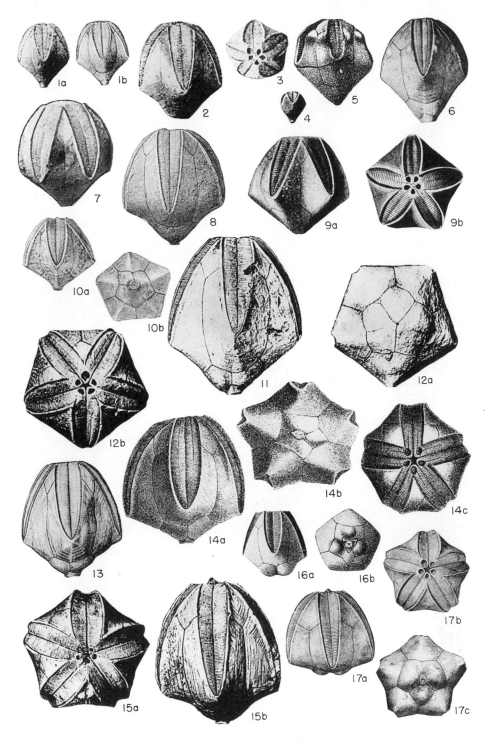

*PENTREMITES SULCATUS* GROUP

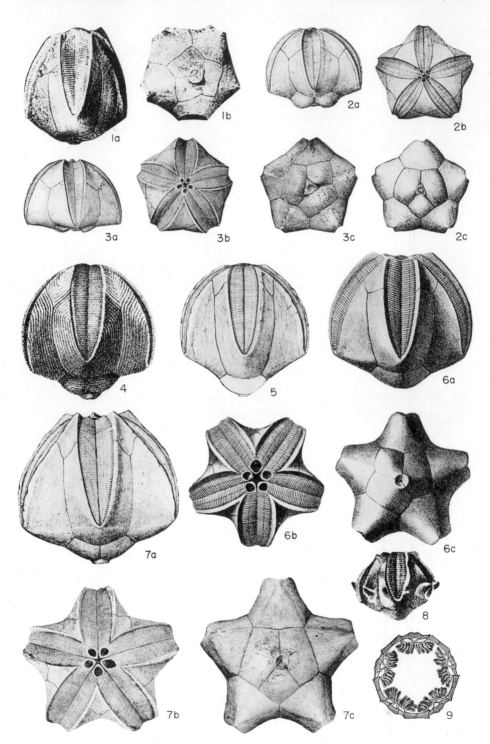

*PENTREMITES SULCATUS* GROUP

PLATE 7.—*PENTREMITES SULCATUS* GROUP

All figures × 1 unless otherwise indicated

Figure                                                             Page

1. *Pentremites laminatus* Easton (1943, Pl. 21, figs. 9, 10). × 2. Upper Chester, Pitkin formation, Leslie, Ark................................................................. 67

2, 3. *Pentremites gutschicki* n. sp. (2) drawing of holotype; (3) photograph of paratype, with base somewhat crushed. Kinkaid formation, Alexander Stone Co. quarry, 8 miles northwest of Marion, Ky. Coll. by R. C. Gutschick. Ind. Univ. Paleo. Coll., Nos. 5255, 5255a. (*See* Pl. 12, fig. 21)................................................................. 68

4. *Pentremites hemisphericus* Hambach (1880, Pl. B, fig. 7). After type figure. Chester limestone, Chester or Evansville, Ill................................................. 68

5. *Pentremites hemisphericus* Hambach. Golconda limestone, railroad just north of Golconda, Ill. Considerably restored. Coll. C. A. Malott. Ind. Univ. Paleo. Coll. No. 5256....... 68

6. *Pentremites sulcatus* (Roemer) (1851, Pl. 6, fig. 10a–c). After type figure. Chester, Prairie du Long, south of Bellville, Ill................................................ 68

7. *Pentremites sulcatus* (Roemer). Beautiful and typical specimen loaned by Dr. Courtney Werner. (a) drawing; (b, c) photographs. Middle or Upper Chester, Evansville, Ill. Washington Univ. Paleo. Coll., No. 820103......................................... 68

8. *Pentremites spinosus* Hambach (1880, Pl. B, fig. 1). After type figure. Chester limestone, Chester, Ill............................................................... 69

9. Idealized cross section of *Pentremites* to show five paired groups of folded hydrospires. After Springer (1913, p. 165). Species cannot be *P. sulcatus*, for ambulacra are convex, not concave. Probably a species of the *P. pyriformis* group........................... 11

93

PLATE 8.—*PENTREMITES SULCATUS* GROUP

All figures × 1 unless otherwise indicated

Figure                                                                                          Page

1. *Pentremites chesterensis* Hambach (1880, Pl. B, fig. 8). After type figure. Chester limestone,
   Randolph Co., or Chester, Ill. . . . . . . . . . . . . . . . . . . . . . . . . . . . . . . . . . . . . . . . . . . . . . . . . . . . 69
2. *Pentremites maccalliei* Schuchert (1906, p. 759, Figs. 1a, b). After type figures. (a) distorted
   specimen slightly tilted to bring out more nearly the normal form; (b) side view, showing
   amount of distortion. Bangor limestone, Nickajack gulch, Cole City, Ga. U. S. Nat. Mus.
   Coll., No. 35689 . . . . . . . . . . . . . . . . . . . . . . . . . . . . . . . . . . . . . . . . . . . . . . . . . . . . . . . . . . . . . 70
3. *Pentremites maccalliei* Schuchert. Golconda formation. Locality unknown, probably Georgia.
   Ind. Univ. Paleo. Coll., No. 4612 . . . . . . . . . . . . . . . . . . . . . . . . . . . . . . . . . . . . . . . . . . . . . . . 70
4. *Pentremites obesus* Lyon (1857, Pl. 2, fig. 1, 1a, 1b). After type figure, reproduced in Lesley
   (1889, p. 621). Golconda limestone, Crittenden Co., Ky. . . . . . . . . . . . . . . . . . . . . . . . . . . . . 70
5. *Pentremites obesus* Lyon. After Hall (1858, Pl. 25, fig. 15). Chester limestone, Southern Ill.
   or Ky. . . . . . . . . . . . . . . . . . . . . . . . . . . . . . . . . . . . . . . . . . . . . . . . . . . . . . . . . . . . . . . . . . . . . . . . 70
6. *Pentremites obesus* Lyon. After type figures of *P. spicatus altipelvis* Haas (1945, p. 2, Figs. 3,
   6). Glen Dean limestone, Crane, Martin Co., Ind. . . . . . . . . . . . . . . . . . . . . . . . . . . . . . . . . . . 70

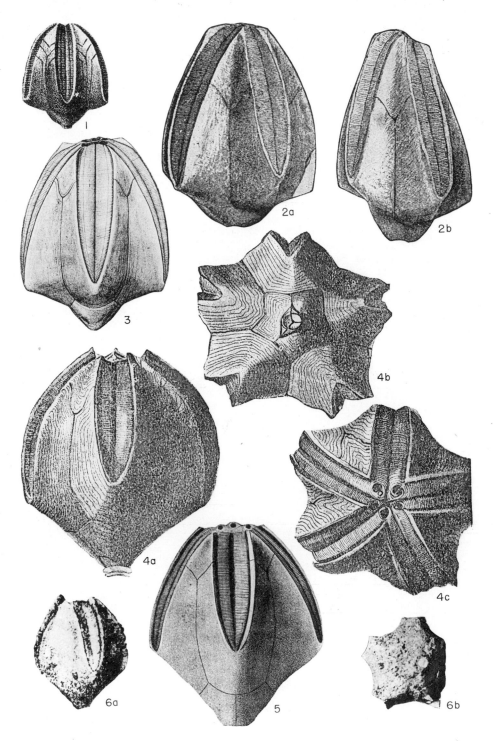

PENTREMITES SULCATUS GROUP

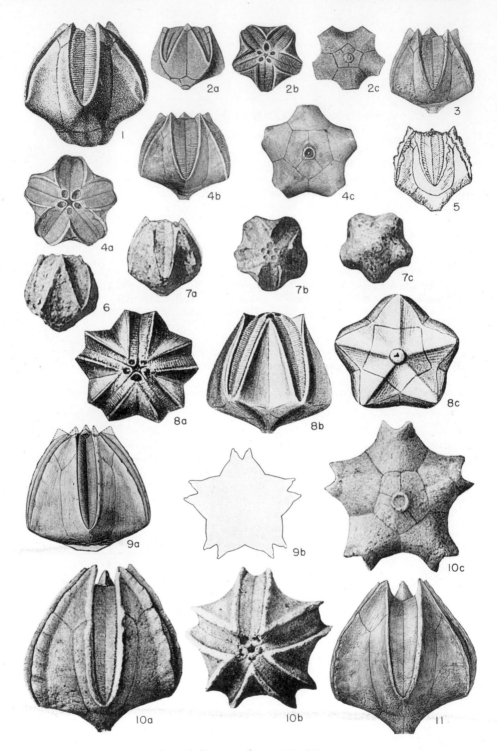

*PENTREMITES SULCATUS* GROUP

PLATE 9.—*PENTREMITES SULCATUS* GROUP

All figures × 1 unless otherwise indicated

Figure                                                                                          Page

1. *Pentremites broadheadi* Hambach (1880, Pl. B, fig. 6). After type figure. Chester, probably Glen Dean formation, Evansville, Ill. . . . . . . . . . . . . . . . . . . . . . . . . . . . . . . . . . . . . . . . . . . . 71

2. *Pentremites halli* n. sp. After Hall's figure of *P. cherokeeus* (not Troost), 1858, Pl. 25, fig. 12a–c. Middle Chester, Chester, Ill. . . . . . . . . . . . . . . . . . . . . . . . . . . . . . . . . . . . . . . 71

3. *Pentremites halli* n. sp. Holotype, somewhat restored. Glen Dean limestone, old quarries 1 mile east of Cloverport, Ky. Ind. Univ. Paleo. Coll., No. 5257 . . . . . . . . . . . . . . . . . . . . 71

4. *Pentremites halli* n. sp. Paratype, partially restored. Same horizon, locality, and depository as (3). No. 5258 . . . . . . . . . . . . . . . . . . . . . . . . . . . . . . . . . . . . . . . . . . . . . . . . . . . . . . . 71

5. *Pentremites halli* n. sp. After *P. cherokeeus* (not Troost but Hall) in Rowley (1903, Pl. 36, fig. 6). Chester limestone, Clifty Sta., Grayson Co., Ky. G. K. Greene Coll. . . . . . . . . . . 71

6. *Pentremites halli* n. sp. After figure of *P. spicatus porrectus* Haas, part (1945, p. 2, Fig. 10). Glen Dean limestone, Crane, Martin Co., Ind. Amer. Mus. Nat. Hist. Coll., No. 26058/2:3 . . . . . . . . . . . . . . . . . . . . . . . . . . . . . . . . . . . . . . . . . . . . . . . . . . . . . . . . . . . 71

7. *Pentremites halli* n. sp. After figure of *P. spicatus altipelvis* Haas, part (1945, p. 2, figs. 1–13). Glen Dean limestone, Crane, Martin Co., Ind. Amer. Mus. Nat. Hist. Coll., No. 26059/2:1. 71

8. *Pentremites cherokeeus* Troost. After type figure in Wood (1909, pl. 3, figs. 14–16). Base of Lookout Mountain, Cherokee Co., Tenn. . . . . . . . . . . . . . . . . . . . . . . . . . . . . . . . . . . . . . 72

9. *Pentremites cherokeeus* Troost. Partially restored. (a) side view (b) outline of cross section to show V-shaped ambulacra. Glen Dean limestone, 1 mile east Cloverport, Ky. Ind. Univ. Paleo. Coll., No. 5259 . . . . . . . . . . . . . . . . . . . . . . . . . . . . . . . . . . . . . . . . . . . . . . . . . 72

10. *Pentremites spicatus* Ulrich (1917, Pl. 7, figs. 33–35). After type figures. Cherty beds of the Glen Dean limestone, Grayson Co., Ky. U. S. Nat. Mus., Springer Coll. . . . . . . . . . . . . . . 72

11. *Pentremites spicatus* Ulrich. Partially restored. Glen Dean limestone, old quarries 1 mile east of Cloverport, Ky. Ind. Univ. Paleo. Coll., No. 5260 . . . . . . . . . . . . . . . . . . . . . . . . . . 72

PLATE 10.—*PENTREMITES SULCATUS* GROUP

All figures × 1 unless otherwise indicated

Figure      Page

1. *Pentremites spicatus* Ulrich. Glen Dean limestone, old quarries, 1 mile east of Cloverport, Ky. Ind. Univ. Paleo. Coll., No. 5261........................................... 72

2. *Pentremites spicatus* Ulrich. From same locality, horizon, and depository as (1) Summit slightly tilted toward observer, emphasizing protruding deltoids, but making base appear shorter than it is. Ind. Univ. Paleo. Coll., No. 5262............................ 72

3. *Pentremites spicatus* Ulrich. After Haas (1945, p. 2, Fig. 8). Glen Dean limestone, Crane, Martin Co., Ind. Amer. Mus. Nat. Hist. Coll., No. 26057/5:1....................... 72

4. *Pentremites spicatus* Ulrich (after *P. obesus* Rowley, 1903, p. 115, Pl. 36, figs. 1–3). Chester, Grayson Springs, Ky........................................................... 72

5. *Pentremites spicatus porrectus* Haas (1945, p. 2, Figs. 2, 5, 7). After type figure. Glen Dean formation, Crane, Martin Co., Ind. Amer. Mus. Nat. Hist. Coll., No. 26058/2:1........ 73

6. *Pentremites spicatus porrectus* Haas. Glen Dean limestone, Crane, Martin Co., Ind. Ind. Univ. Paleo. Coll., No. 5264............................................................ 73

7. *Pentremites serratus* Hambach (1903, Pl. 4, fig. 9). After type figure. Ste. Genevieve, Mo.... 73

8. *Pentremites serratus* Hambach. Glen Dean limestone, old quarries 1 mile east of Cloverport, Ky. Ind. Univ. Paleo. Coll., No. 5263......................................... 73

9. *Pentremites basilaris* Hambach (1880, Pl. B, fig. 9). After type figure. Chester limestone, Evansville, Ill............................................................... 74

10. *Pentremites basilaris* Hambach. Basal view. × 2/3. Glen Dean limestone, Crane, Martin Co., Ind. Ind. Univ. Paleo. Coll., No. 5265.......................................... 74

11. *Pentremites bradleyi* Meek. After type figure in Hambach (1903, Pl. 5, fig. 7). Subcarboniferous, Mont. Smithsonian Inst. Coll., No. 24529....................................... 74

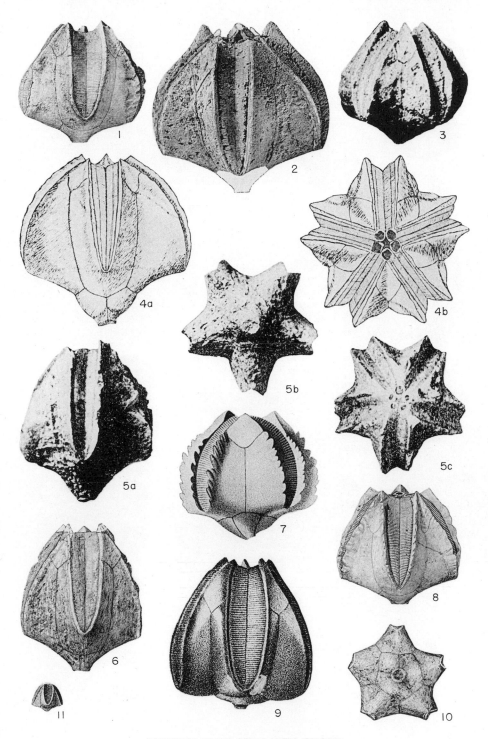

*PENTREMITES SULCATUS* GROUP

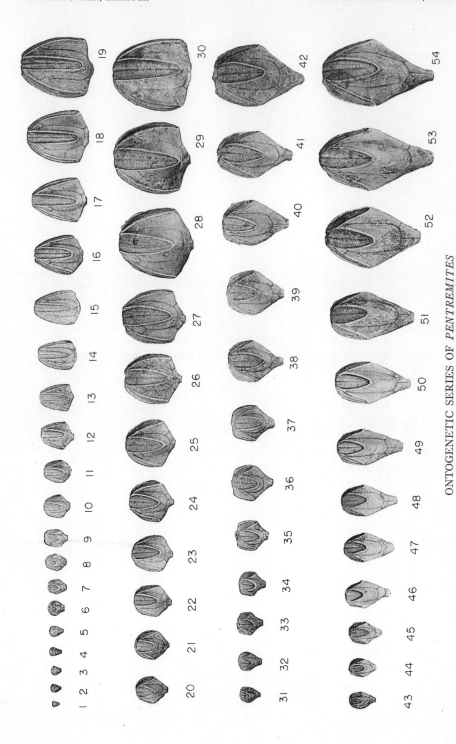

ONTOGENETIC SERIES OF *PENTREMITES*

PLATE 11.—ONTOGENETIC SERIES OF *PENTREMITES*

All figures natural size

Figures                                                                                          Page

1–19. *Pentremites conoideus* Hall. Note enlargement of pelvic angle and lengthening of vault during ontogeny. (*See* Pl. 1, figs. 11–13, for enlarged drawings of first, third, and eighth specimens from left.) Specimens 1–5 are nepionic; 6–11 are neanic; 12–19 are ephebic. Part of series published by Smith (1906, Pl. 46, figs. 1, 2). Upper Harrodsburg limestone, "Pentremite Hollow", 2 miles south of Bloomington; Salem limestone, old Cleveland quarry, Harrodsburg, Ind., and Matthews quarry, Ellettsville, Ind....... 42

20–30. *Pentremites godoni* (Defrance). Specimens 20–24 neanic; 25–30 ephebic. Neanic stages have smaller pelvic angles than ephebic specimens and tend to be globose, as in early neanic stages of most *Pentremites*, causing misidentification as "*P. globosus* Say", which is unrecognizable. Paint Creek formation, Prairie du Long Creek, near Floraville, St. Clair Co., Ill. ................................................................. 48

31–42. *Pentremites girtyi* Ulrich. Specimens 31–37 neanic; 38–42 ephebic. Small pelvic angle, about 60°, retained throughout ontogeny, but pelvis elongates more than vault. Golconda limestone, Beaver Creek, half a mile southeast of Shoals, Ind................ 59

43–54. *Pentremites okawensis* Weller. Specimens 43–45 ananeanic; 46–48, metaneanic; 49–51, paraneanic; 52–54, ephebic. Pelvic angle of about 50° retained during ontogeny, vault lengthens relatively more than pelvis, and form ratio decreases. Golconda limestone, Beaver Creek, half a mile southeast of Shoals, Ind................................. 60

PLATE 12.—ONTOGENIC SERIES OF *PENTREMITES*

All figures natural size

Figure                                                                    Page

1-12. *Pentremites sulcatus* (Roemer). Ontogenetic series after Hambach (1903, Pl. 6, figs. 1–12).
Note enlargement of pelvic angle during ontogeny. (1–8) early to late neanic specimens, in *P. angularis* ancestral stage; (9–11) typical ephebic specimens; (12) gerontic
specimen almost at *P. spicatus* stage. Hambach does not give horizon or locality, but
states that all specimens are from same locality. Age probably Glen Dean . . . . . . . . . . . 68

13-21. *Pentremites gutschicki* n. sp. (13–17) neanic specimens in *P. tulipaformis* stage, *i.e.*, ancestral stage; (18–21) ephebic specimens; (21) retouched photograph of holotype, illustrated by drawing, Figures 2a–c of Plate 7. Kinkaid limestone, 8 miles, northeast of
Marion, Ky. Coll. by R. C. Gutschick . . . . . . . . . . . . . . . . . . . . . . . . . . . . . . . . . . . . . . . . . 68

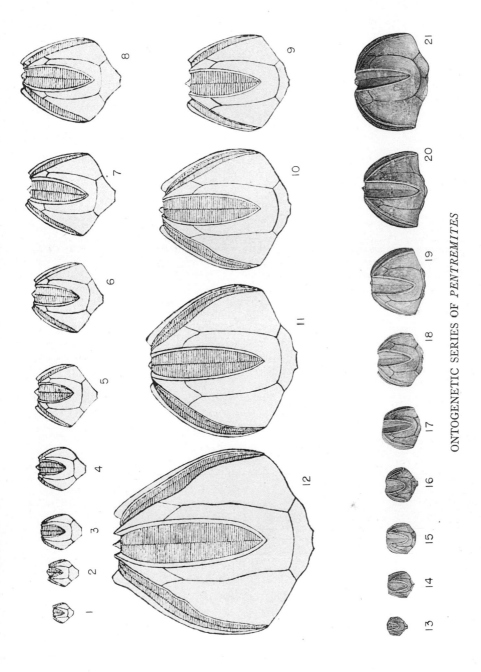

ONTOGENETIC SERIES OF *PENTREMITES*

*PENTREMITES*

PLATE 13.—STRUCTURES AND FOSSILIZATION OF *PENTREMITES*

Figure                                                           Page

1. *Pentremites godoni angustus* Hambach. × 1⅓. Showing growth lines. Growth in radials from base of ambulacra, upward, outward, and downward; downward from upper tips of deltoids; concentrically from centers of basals; upward from tips of ambulacra. (*See* Pl. 1, fig. 3.) Paint Creek formation, Prairie du Long Creek near Floraville, Ill. Ind. Univ. Paleo. Coll., No. 5266.................................................. 50

2. *Pentremites pyriformis* Say? After figure of *Pentatrematites sulcatus* Roemer (part) (1851, Pl. 6, fig. 7a) to show pinnules in place. × 1..................................... 56

3. *Pentremites pyriformis* Say, of Etheridge and Carpenter (1886, Pl. 1, fig. 2, part), showing finer characters of ambulacrum. × 10. No *Pentremites* studied show lateral pores with lower rims, or narrow comma-shaped grooves on ridges......................... 56

4. *Pentremites halli* n. sp. (a) × 6. Middle part of ambulacrum. Pits at sutures between side plates and lancet plate unusual for this or any other species; 8–9 transverse ridges in 3 mm; (b–d) × 1. Retouched photographs, showing flaring, platelike deltoids; basal angle 115°, 10° less than normal for species. Glen Dean limestone, 2 miles southeast of Herod, Ill. Ind. Univ. Paleo. Coll., No. 5269a................................... 71

5. *Pentremites halli* n. sp. × 2. Half-grown specimen, 11 mm long and wide, showing concave ambulacra, flaring deltoids which scarcely reach summit; 9 transverse ridges in 3 mm; basal angle 110°. Apertural view appears normal; in basal view unpaired plate is left posterior instead of right anterior; only specimen seen with unpaired plate not in usual right anterior position. Glen Dean, 2 miles southeast of Herod, Ill. Ind. Univ. Paleo. Coll., No. 5269b....................................................... 71

6. Upper part of ambulacrum of *P. fohsi* Ulrich (1905, Pl. 7, fig. 9). × 3. Ambulacrum concave, transverse ridges not inclined, normal for large specimens of any species of *P. sulcatus* group; 9 transverse ridges in 3 mm. Probably middle Chester, 8 miles northwest of Princeton, Ky........................................................................ 66

7. Part of ambulacrum of *P. pyramidatus* Ulrich. × 3. (After Ulrich, 1905, Pl. 7, fig. 14.) Grooves on ridges extend onto lancet plate. Only 6 transverse ridges in lower part of ambulacrum, increasing to 9 or 10 in upper part (not shown). Probably middle Chester, western Kentucky............................................................................ 57

8. *Pentremites conoideus* Hall. × 1. Crushed specimen showing stem in place; stem with spines. Salem limestone, West Point, Ky. Ind. Univ. Paleo. Coll., No. 4642........... 42

9–12. *Pentremites godoni* (Defrance). × 1. Oversilicified specimens just at geode stage. (9–11) apertural views showing ambulacra and polygonal areas bounded by quartz veins, making welts. Specimens 1.5 to 1.7 times original diameters. (12) Basal view showing tips of ambulacra and stem and polygonal areas of quartz veins. Original shell and veins now chalcedony, interior filled with quartz, leaving small central cavity. Lower or middle Chester, Warren Co., Ky. Ind. Univ. Paleo. Coll., No. 3135............... 48

# INDEX*

Abnormalities, 24
Abstract, 1
Abundance, 25
Allen, A. T., and Lester, J. G., 33, 70
Ambulacra, 10, 11, 26
   convexity or concavity, 14
   fine structures, 18
   length, 18
   width, 17
Ambulacral flange, 11, 15, 27
Ambulacral ratio, 11, 18
Ambulacral rim, 11, 15, 27
Anal opening, 10, 16
Anal pore, 11
Ananepionic stage, 29
Ancestry of *Pentremites*, 30
Angle between pelvis and vault, 20
Angle of vault, 20
Apex, 11
Axis, 13

Basal angle, 11, 14
Basal periphery, 11
Basalia, 11
Basal plates, 10, 11
   shape of, 10, 23
Base, 11
Beede, J. W., 43
Billings, E., 24
Brachioles, 11
Burma, B. H., 13
Butts, C., 3, 10, 24, 35, 43, 70

Calyx, 8, 11
Casts, 35
Characteristics of *Pentremites*, 12
Chester Series, 9
Clark, T. H., 3, 13, 25, 43, 49
Classification of Meramec-Chester formations, 9
Collecting localities, 75
Columnals, 8, 11
Concavity of ambulacra, 14
Concavity of interambulacra, 15
Conrad, T. A., 3
Convexity of ambulacra, 14
Croneis, C., and Geis, H. L., 30
Crushed specimens, 35
Crystallization, 34

---

*Figures in **boldface** refer to descriptions.

Definition of a pentremite, **8**
Defrance, J. L. M., 3, 5
Deltoid plates, 8, 11, 27
   basal suture, 23
   length, 20
Distortion, 35
Dorsad, 11
Dorsal region, 11
Dwarfing, 21

Easton, W. H., 3, 24, 68
Ecology of *Pentremites*, 32
Elrod, M. N. 70
*Encrina*, 37
   *Godonii*, 5, 37, 48
*Encrinites*, 37, 48
   *florealis*, 48
*Encrinus*, 6
Ephebic stage, 11, 29
Etheridge, R., Jr., and Carpenter, P. H., 3, 7, 10, 18, 24, 27, 35, 45, 75
Etymology of name *Pentremites*, 6
Evolution, 31
Extension of deltoids, 15

Fine structures of ambulacra, 18
Flare of deltoids, 16
Forked plates, 8, 11
Fossilization of *Pentremites*, 33

Geodization, 34
Geographic occurrence, 25
Geological history of *Pentremites*, 25
Geologic occurrence, 28, Fig. 1
   *P. conoideus* group, 40, Fig. 2
   *P. godoni* group, 46, Fig. 3
   *P. pyriformis* group, 52, Fig. 4
   *P. sulcatus* group, 62, Fig. 5
Gerontic stage, 11
Glossary, 11
Goldfuss, A., 6
Greger, D. K., 3, 8
Groups of *Pentremites*, 26, 31, 32, 38
Growth, 10, 29, Pl. 13, fig. 1
Guide for description of species, 38
Gutschick, R. C., 55, 67

Haas, O., 3, 4, 69, 73
Hall, J., 4, 7, 13, 21, 57, 66, 71, 72, 74
Hambach, G., 3, 7, 10, 13, 18, 21–24, 27, 29, 30, 59, 68, 69, 72, 73, 76

101

Height of ambulacral rim, 15
Historical review and nomenclature of *Pentremites*, 5
Hydrospire pores, 11
Hydrospires, 10, 11, 24

Infiltration, 33
Interambulacral areas, 11, 26
Interambulacral margins, 11
Internal molds, 35
Introduction, 3

Kentucky Asterial fossil, 6
Key to groups of *Pentremites*, 39
Keys, C. R., 27, 43
Keys to species, 39, 45, 51, 61

Lancet plate, 10, 11
    width of, 22
Lateral grooves, 11
    number in 3 mm, 18
    slope of, 18
Lateral pores, 12, 22
Length of ambulacra, 12, 18
Length of calyx, 12
Length of deltoid plates, 20
Length of stem, 23
Length to width ratio or L/W, 12, 13
Lyon, S. S., 3, 7, 10, 22, 23, 76

Malott, C. A., v, 24, 76
Mather, K. F., 26, 50
Median groove, 10, 22
Meek, F. B., 3
Meek, F. B., and Worthen, A. H., 3
Meramec series, 9
Metanepionic stage, 29
Miller, S. A., 6
Minute ridges and grooves, 19
Mitchill, S. L., 5
Morgan, G. O., 26
Most important characters, 13
Mouth, 10, 12
    shape and size, 22

Neanic stage, 12
Negligible characters, 21
Nepionic stage, 12, 21
Nodes on basal and radial plates, 16
Nonstructural features, 25
Number of lateral grooves in 3 mm, 18

Oblique band, 19
Ontogenetic development, 27

Ontogenetic series, Pls. 11, 12
Ontogenetic stages, 21, 29
Orientation, 13
Outside plates, 12
Owen, D. D., 59

Paranepionic stage, 29
Parkinson, J., 5
Peck, R. E., 43
Pelvic angle, 12, 14, 32
Pelvic ridges, 23
Pelvis, 12
*Pentatrematites*, 37
    *florealis*, 48
    *sulcatus*, 68
*Pentatremites*, 37, 48
    *florealis*, 48
    *ovalis*, 44
*Pentetrematites*, 79
*Pentremite*, 6, 8, 37
    *florealis*, 48
*Pentremites*, 3, 5, 6, 8, 10, 11, 12, 14, 17, 21, 24, 25–35, 37
    *abbreviatus*, 49
    *abruptus*, 54
    *abruptus* var., 56
    *altimarginatus*, 47
    *altus*, 55
    *angularis*, **64**
    *angustus*, 50
    *arctibrachiatus*, 56
    *arctibrachiatus huntsvillensis*, 56
    *basilaris*, **74**
    *benedicti*, 42
    *biconvexus*, **47**
    *bradleyi*, **74**
    *brevis*, 49, 67
    *broadheadi*, **71**
    *burlingtonensis*, **42**
    *buttsi*, **58**
    *calycinus*, **64**
    *canalis*, 64
    *cavus*, 44
    *cervinus*, **66**
    *cherokeeus*, **72**
    *chesterensis*, **69**
    *clavatus*, **60**
    *conoideus*, 27, 30, **42**
    *conoideus amplus*, 44
    *conoideus* group, 26, 31, 32, 39, **41**
        Key, 39
        Main characters, 41
    *conoideus obtusus*, **43**
    *conoideus perlongus*, 30, **43**

*decipiens*, 55
*decipiens decurtatus*, 55
*decussatus*, **41**
*divergens*, 54
*downeyensis*, 43
*elegans*, **64**
*elongatus*, **42**
*florealis*, 48
*fohsi*, **66**
*fohsi marionensis*, 66
*gemmiformis*, **58**
*gianteus*, 70
*girtyi*, **59**
*globosa*, *-us*, 74
*godoni*, 48
*godoni abbreviatus*, **49**
*godoni angustus*, 50
*godoni* group, 26, 31 32, **39**
    Key, 45
    Main characters, 47
*godoni major*, 30, **49**
*godoni pinguis*, **45**
*grandis*, **75**
*gutschicki* n. sp., 68
*halli* n. sp., **71**
*hambachi*, 65
*hemisphericus*, **68**
*kirki*, **58**
*koninckana*, *-us*, 42
*laminatus*, **67**
*leda*, 40
*lyoni*, 56
*lyoni gracilens*, 58
*maccalliei*, 33, **70**
*malotti* n. sp., **48**
*nodosus*, **65**
*obesus*, **70**, 72
*obtusus*, 43
*okawensis*, **60**
*ovalis*, **44**
*ovoides*, **44**
*patei*, 57
*pediculatus*, 56
*perelongatus*, **75**
*pinguis*, 45
*planus*, 49
*platybasis*, **50**
*prematurus*, 65
*princetonensis*, **52**
*pulchellus*, **51**
*pyramidatus*, **57**
*pyriformis*, **56**, 58
*pyriformis* group, 26, 31, 32, **39**

    Key, 51
    Main characters, 53
*robustus*, **66**
*rusticus*, 50
*saximontanus*, 55
*serratus*, **73**
*speciosus*, 59
*spicatus*, **72**
*spicatus altipelvis*, 70, 71
*spicatus porrectus*, 71, **73**
*spinosus*, **69**
*springeri*, **62**
*sulcatus*, **68**, 72
    Key, 61
    Main characters, 63
*symmetricus*, **55**
*tulipaformis*, **67**
*turbinatus*, 59
*tuscumbiae*, **46**
*welleri*, **56**
*Pentremitidea*, 10, 11, 40
    *leda*, 40
Phylogeny of groups of *Pentremites*, 38, Fig. 1
    *P. conoideus* group, 40, Fig. 2
    *P. godoni* group, 46, Fig. 3
    *P. pyriformis* group, 52, Fig. 4
    *P. sulcatus* group, 62, Fig. 5
Phylogeny of *Pentremites*, 31
Pinnules, 10, 12, 24
Pits, 10, 12, 19
Poral pieces, 11, 12
Pores, 12
Position of greatest width, 17
Profile of calyx, 12
Profile of edges of pelvis, 16

Radial pelvic ridges, 23
Radial plates, 8, 12
Radial sinuses, 12
Roemer, F., 3, 6, 10, 24, 27
Rowley, R. R., 3, 4, 7, 21, 43, 67, 70, 71, 76

Say, T., 3, 6, 74
Schlotheim, E. F. von, 3, 6
Schuchert, C., 3
Secondary characters, 17
Secondary grooves, 19
Shape of sides of vault, 16
Shape of specimen, 17
Shape of summit, 20
Shumard, B. F., 3, 7, 22, 24
Side plates, 11, 12
    width of, 22
Silicification, 34

Size of specimen, 17
Slope of lateral grooves, 18, 22
Smith, E. A., 21, 27, 29, 30, 43
Sowerby, G. B., 6
Spiracles, 10, 12
  shape and size, 22
Stem, 8, 12
  length of, 23
  diameter of, 23
Stratigraphic occurrence, 25
Striations on plates, 24
Structure of *Pentremites*, 8
Summit, 10, 12
  plates, 24
  shape of, 20
  width of, 20
Sutures  8, 12, 23
Swallow, G. C., 3
Systematic descriptions, 37

Theca, 12
Transverse grooves, 10, 11, 12
  number in 3 mm, 18
  slope of grooves, 18
  total number, 21
Transverse ridges, 10, 12, 19, 21

Troost  G., 3, 25, 72, 74
Twenhofel, W. H., and Shrock, R. R., 27

Ulrich, E. O., 3, 4, 8, 13, 18, 19, 21, 25, 31, 33, 43,
  55-59, 62, 65, 76
Ulrich, E. O., Smith, W. S. T., 71, 76

Van Tuyl, F. M., 43
Variation, 4
Vault, 12, 16
  angle, 20
  profile, 16
Vault to pelvis ratio or V/P, 12, 13
Ventrad, 12
Ventral region, 12

Warren, P. S., 26
Weller, J. M., 8
Weller, J. M., and Sutton, A. H., 15, 71
Weller, S., 3, 5, 8, 13, 24, 25, 57, 65, 68, 73, 76
Werner, C., 22, 24, 69
Whitfield, R. P., 21, 43
Width of ambulacra, 12, 17
Width of calyx, 12, 17
Width of lancet and side plates, 22
Width of summit, 20
Wood, E., 72, 73, 74